Trigonometric Delights

Trigonometric Delights

Eli Maor

PRINCETON UNIVERSITY PRESS • PRINCETON, NEW JERSEY

Copyright © 1998 by Princeton University Press
Published by Princeton University Press, 41 William Street,
Princeton, New Jersey 08540
In the United Kingdom: Princeton University Press, Chichester,
West Sussex

Maor, Eli.
    Trigonometric delights / Eli Maor.
        p.   cm.
    Includes bibliographical references and index.
    ISBN 0-691-05754-0 (alk. paper)
    1. Trigonometry.   I. Title.
    QA531.M394   1998
    516.24'2—dc21   97-18001

This book has been composed in Times Roman

Princeton University Press books are printed on acid-free paper and
meet the guidelines for permanence and durability of the Committee
on Production Guidelines for Book Longevity of the Council on
Library Resources

Printed in the United States of America

10 9 8 7 6 5 4 3

http://pup.princeton.edu

In memory of my uncles

Ernst C. Stiefel (1907–1997)

Rudy C. Stiefel (1917–1989)

Contents

Title page of the Rhind Papyrus.

Preface

There is perhaps nothing which so occupies the middle
position of mathematics as trigonometry.
—J. F. HERBART (1890)

This book is neither a textbook of trigonometry—of which there
are many—nor a comprehensive history of the subject, of which
there is almost none. It is an attempt to present selected topics
in trigonometry from a historic point of view and to show their
relevance to other sciences. It grew out of my love affair with
the subject, but also out of my frustration at the way it is being
taught in our colleges.

First, the love affair. In the junior year of my high school we
were fortunate to have an excellent teacher, a young, vigorous
man who taught us both mathematics and physics. He was a
no-nonsense teacher, and a very demanding one. He would not
tolerate your arriving late to class or missing an exam—and you
better made sure you didn't, lest it was reflected on your report
card. Worse would come if you failed to do your homework or
did poorly on a test. We feared him, trembled when he repri-
manded us, and were scared that he would contact our parents.
Yet we revered him, and he became a role model to many of
us. Above all, he showed us the relevance of mathematics to
the real world—especially to physics. And that meant learning
a good deal of trigonometry.

He and I have kept a lively correspondence for many years,
and we have met several times. He was very opinionated, and
whatever you said about any subject–mathematical or otherwise—he would argue with you, and usually prevail. Years after I finished my university studies, he would let me understand that *he* was still my teacher. Born in China to a family
that fled Europe before World War II, he emigrated to Israel
and began his education at the Hebrew University of Jerusalem,
only to be drafted into the army during Israel's war of indepen-
dence. Later he joined the faculty of Tel Aviv University and
was granted tenure despite not having a Ph.D.—one of only two
faculty members so honored. In 1989, while giving his weekly

lecture on the history of mathematics, he suddenly collapsed and died instantly. His name was Nathan Elioseph. I miss him dearly.

And now the frustration. In the late 1950s, following the early Soviet successes in space (Sputnik I was launched on October 4, 1957; I remember the date—it was my twentieth birthday) there was a call for revamping our entire educational system, especially science education. New ideas and new programs suddenly proliferated, all designed to close the perceived technological gap between us and the Soviets (some dared to question whether the gap really existed, but their voices were swept aside in the general frenzy). These were the golden years of American science education. If you had some novel idea about how to teach a subject—and often you didn't even need that much—you were almost guaranteed a grant to work on it. Thus was born the "New Math"—an attempt to make students *understand* what they were doing, rather than subject them to rote learning and memorization, as had been done for generations. An enormous amount of time and money was spent on developing new ways of teaching math, with emphasis on abstract concepts such as set theory, functions (defined as sets of ordered pairs), and formal logic. Seminars, workshops, new curricula, and new texts were organized in haste, with hundreds of educators disseminating the new ideas to thousands of bewildered teachers and parents. Others traveled abroad to spread the new gospel in developing countries whose populations could barely read and write.

Today, from a distance of four decades, most educators agree that the New Math did more harm than good. Our students may have been taught the language and symbols of set theory, but when it comes to the simplest numerical calculations they stumble—with or without a calculator. Consequently, many high school graduates are lacking basic algebraic skills, and, not surprisingly, some 50 percent of them fail their first college-level calculus course. Colleges and universities are spending vast resources on remedial programs (usually made more palatable by giving them some euphemistic title like "developmental program" or "math lab"), with success rates that are moderate at best.

Two of the casualties of the New Math were geometry and trigonometry. A subject of crucial importance in science and engineering, trigonometry fell victim to the call for change. Formal definitions and legalistic verbosity—all in the name of mathematical rigor—replaced a real understanding of the subject. Instead of an angle, one now talks of the measure of an angle; instead of defining the sine and cosine in a geometric context—

as ratios of sides in a triangle or as projections of the unit circle on the x- and y-axes—one talks about the wrapping function from the reals to the interval $[-1, 1]$. Set notation and set language have pervaded all discussion, with the result that a relatively simple subject became obscured in meaningless formalism.

Worse, because so many high school graduates are lacking basic algebraic skills, the level and depth of the typical trigonometry textbook have steadily declined. Examples and exercises are often of the simplest and most routine kind, requiring hardly anything more than the memorization of a few basic formulas. Like the notorious "word problems" of algebra, most of these exercises are dull and uninspiring, leaving the student with a feeling of "so what?" Hardly ever are students given a chance to cope with a really challenging identity, one that might leave them with a sense of accomplishment. For example,

1. Prove that for any number x,

$$\frac{\sin x}{x} = \cos \frac{x}{2} \cos \frac{x}{4} \cos \frac{x}{8} \cdots .$$

This formula was discovered by Euler. Substituting $x = \pi/2$, using the fact that $\cos \pi/4 = \sqrt{2}/2$ and repeatedly applying the half-angle formula for the cosine, we get the beautiful formula

$$\frac{2}{\pi} = \frac{\sqrt{2}}{2} \cdot \frac{\sqrt{2+\sqrt{2}}}{2} \cdot \frac{\sqrt{2+\sqrt{2+\sqrt{2}}}}{2} \cdots ,$$

discovered in 1593 by François Viète in a purely geometric way.

2. Prove that in any triangle,

$$\sin \alpha + \sin \beta + \sin \gamma = 4 \cos \frac{\alpha}{2} \cos \frac{\beta}{2} \cos \frac{\gamma}{2},$$

$$\sin 2\alpha + \sin 2\beta + \sin 2\gamma = 4 \sin \alpha \sin \beta \sin \gamma,$$

$$\sin 3\alpha + \sin 3\beta + \sin 3\gamma = -4 \cos \frac{3\alpha}{2} \cos \frac{3\beta}{2} \cos \frac{3\gamma}{2},$$

$$\tan \alpha + \tan \beta + \tan \gamma = \tan \alpha \tan \beta \tan \gamma.$$

(The last formula has some unexpected consequences, which we will discuss in chapter 12.) These formulas are remarkable for their symmetry; one might even call them "beautiful"—a kind word for a subject that has undeservedly gained a reputation of being dry and technical. In Appendix 3, I have collected some additional beautiful formulas, recognizing of course that "beauty" is an entirely subjective trait.

"Some students," said Edna Kramer in *The Nature and Growth of Modern Mathematics*, consider trigonometry "a glorified geometry with superimposed computational torture." The present book is an attempt to dispel this view. I have adopted a historical approach, partly because I believe it can go a long way to endear mathematics–and science in general—to the students. However, I have avoided a strict chronological presentation of topics, selecting them instead for their aesthetic appeal or their relevance to other sciences. Naturally, my choice of subjects reflects my own preferences; numerous other topics could have been selected.

The first nine chapters require only basic algebra and trigonometry; the remaining chapters rely on some knowledge of calculus (no higher than Calculus II). Much of the material should thus be accessible to high school and college students. Having this audience in mind, I limited the discussion to plane trigonometry, avoiding spherical trigonometry altogether (although historically it was the latter that dominated the subject at first). Some additional historical material–often biographical in nature—is included in eight "sidebars" that can be read independently of the main chapters. If even a few readers will be inspired by these chapters, I will consider myself rewarded.

My dearest thanks go to my son Eyal for preparing the illustrations; to William Dunham of Muhlenberg College in Allentown, Pennsylvania, and Paul J. Nahin of the University of New Hampshire for their very thorough reading of the manuscript; to the staff of Princeton University Press for their meticulous care in preparing the work for print; to the Skokie Public Library, whose staff greatly helped me in locating rare and out-of-print sources; and last but not least to my dear wife Dalia for constantly encouraging me to see the work through. Without their help, this book would have never seen the light of day.

Note: frequent reference is made throughout this book to the *Dictionary of Scientific Biography* (16 vols.; Charles Coulston Gillispie, ed.; New York: Charles Scribner's Sons, 1970–1980). To avoid repetition, this work will be referred to as *DSB*.

Skokie, Illinois
February 20, 1997

Trigonometric Delights

PROLOGUE

Ahmes the Scribe, 1650 B.C.

Soldiers: from the summit of yonder pyramids forty
centuries look down upon you.
—NAPOLEON BONAPARTE in Egypt, July 21, 1798

In 1858 a Scottish lawyer and antiquarian, A. Henry Rhind
(1833–1863), on one of his trips to the Nile valley, purchased a
document that had been found a few years earlier in the ruins
of a small building in Thebes (near present-day Luxor) in Upper Egypt. The document, known since as the Rhind Papyrus,
turned out to be a collection of 84 mathematical problems dealing with arithmetic, primitive algebra, and geometry.[1] After
Rhind's untimely death at the age of thirty, it came into the
possession of the British Museum, where it is now permanently
displayed. The papyrus as originally found was in the form of
a scroll 18 feet long and 13 inches wide, but when the British
Museum acquired it some fragments were missing. By a stroke
of extraordinary luck these were later found in the possession
of the New-York Historical Society, so that the complete text is
now available again.

Ancient Egypt, with its legendary shrines and treasures,
has always captivated the imagination of European travelers.
Napoleon's military campaign in Egypt in 1799, despite its
ultimate failure, opened the country to an army of scholars, an-
tiquarians, and adventurers. Napoleon had a deep interest in
culture and science and included on his staff a number of scholars in various fields, among them the mathematician Joseph
Fourier (about whom we will have more to say later). These
scholars combed the country for ancient treasures, taking with
them back to Europe whatever they could lay their hands on.
Their most famous find was a large basalt slab unearthed near
the town of Rashid—known to Europeans as Rosetta—at the
western extremity of the Nile Delta.

The Rosetta Stone, which like the Rhind Papyrus ended up
in the British Museum, carries a decree issued by a council

of Egyptian priests during the reign of Ptolemy V (195 B.C.) and is recorded in three languages: Greek, demotic, and hieroglyphic (picture script). The English physicist Thomas Young (1773–1829), a man of many interests who is best known for his wave theory of light, was the first to decipher the inscription on the stone. By comparing the recurrence of similar groups of signs in the three scripts, he was able to compile a primitive dictionary of ancient Egyptian words. His work was completed in 1822 by the famous French Egyptologist, Jean François Champollion (1790–1832), who identified the name Cleopatra in the inscription. Champollion's epochal work enabled scholars to decipher numerous Egyptian texts written on papyri, wood, and stone, among them several scrolls dealing with mathematics. The longest and most complete of the mathematical texts is the Rhind Papyrus.

August Eisenlohr, a German scholar, was the first to translate the Rhind Papyrus into a modern language (Leipzig, 1877); an English translation by Thomas Eric Peet appeared in London in 1923.[2] But the most extensive edition of the work was completed in 1929 by Arnold Buffum Chase (1845–1932), an American businessman whose trip to Egypt in 1910 turned him into an Egyptologist. It is through this edition that the Rhind Papyrus became accessible to the general public.[3]

The papyrus is written from right to left in hieratic (cursive) script, as opposed to the earlier hieroglyphic or pictorial script. The text is in two colors—black and red—and is accompanied by drawings of geometric shapes. It is written in the hand of a scribe named A'h-mose, commonly known to modern writers as Ahmes. But it is not his own work; he copied it from an older manuscript, as we know from his own introduction:

This book was copied in the year 33, in the fourth month of the inundation season, under the majesty of the king of Upper and Lower Egypt, 'A-user-Re', endowed with life, in likeness to writings of old made in the time of the king of Upper and Lower Egypt, Ne-ma'et-Re'. It is the scribe A'h-mose who copies this writing.[4]

The first king mentioned, 'A-user-Re', has been identified as a member of the Hyksos dynasty who lived around 1650 B.C.; the second king, Ne-ma'et-Re', was Amenem-het III, who reigned from 1849 to 1801 B.C. during what is known as the Middle Kingdom. Thus we can fix the dates of both the original work and its copy with remarkable accuracy: it was written nearly four thousand years ago and is one of the earliest, and by far the most extensive, ancient mathematical document known to us.[5]

The work opens with a grand vision of what the author plans to offer: a "complete and thorough study of all things, insight into all that exists, knowledge of all secrets."[6] Even if these promises are not quite fulfilled, the work gives us an invaluable insight into early Egyptian mathematics. Its 84 problems deal with arithmetic, verbal algebra (finding an unknown quantity), mensuration (area and volume calculations), and even arithmetic and geometric progressions. To anyone accustomed to the formal structure of Greek mathematics—definitions, axioms, theorems, and proofs—the content of the Rhind Papyrus must come as a disappointment: there are no general rules that apply to an entire *class* of problems, nor are the results derived logically from previously established facts. Instead, the problems are in the nature of specific examples using particular numbers. Mostly they are "story problems" dealing with such mundane matters as finding the area of a field or the volume of a granary, or how to divide a number of loaves of bread among so many men. Apparently the work was intended as a collection of exercises for use in a school of scribes, for it was the class of royal scribes to whom all literary tasks were assigned—reading, writing, and arithmetic, our modern "three R's."[7] The papyrus even contains a recreational problem of no apparent practical use, obviously meant to challenge and entertain the reader (see p. 11).

The work begins with two tables: a division table of 2 by all odd integers from 3 to 101, and a division table of the integers 1 through 9 by 10. The answers are given in *unit fractions*—fractions whose numerator is 1. For some reason this was the only way the Egyptians knew of handling fractions; the one exception was 2/3, which was regarded as a fraction in its own right. A great amount of effort and ingenuity was spent in decomposing a fraction into a sum of unit fractions. For example, the result of dividing 6 by 10 is given as $1/2 + 1/10$, and that of 7 by 10 as $2/3 + 1/30$.[8] The Egyptians, of course, did not use our modern notation for fractions; they indicated the reciprocal of an integer by placing a dot (or an oval in hieroglyphic script) over the symbol for that integer. There was no symbol for addition; the unit fractions were simply written next to each other, their summation being implied.[9]

The work next deals with arithmetic problems involving subtraction (called "completion") and multiplication, and problems where an unknown quantity is sought; these are known as *aha* problems because they often begin with the word "h" (pronounced "aha" or "hau"), which probably means "the quantity" (to be found).[10] For example, Problem 30 asks: "If the scribe says, What is the quantity of which $2/3 + 1/10$ will make 10, let him hear." The answer is given as $13 + 1/23$, followed by a proof (today we would say a "check") that this is indeed the correct answer.

In modern terms, Problem 30 amounts to solving the equation $(2/3 + 1/10)x = 10$. Linear equations of this kind were solved by the so-called "rule of false position": assume some convenient value for x, say 30, and substitute it in the equation; the left side then becomes 23, instead of the required 10. Since 23 must be multiplied by $10/23$ to get 10, the correct solution will be $10/23$ times the assumed value, that is, $x = 300/23 = 13 + 1/23$. Thus, some 3,500 years before the creation of modern symbolic algebra, the Egyptians were already in possession of a method that allowed them, in effect, to solve linear equations.[11]

Problems 41 through 60 are geometric in nature. Problem 41 simply says: "Find the volume of a cylindrical granary of diameter 9 and height 10." The solution follows: "Take away 1/9 of 9, namely, 1; the remainder is 8. Multiply 8 times 8; it makes 64. Multiply 64 times 10; it makes 640 cubed cubits." (The author then multiplies this result by 15/2 to convert it to *hekat*, the standard unit of volume used for measuring grain; one hekat has been determined to equal 292.24 cubic inches or 4.789 liters.)[12] Thus, to find the area of the circular base, the scribe replaced it by a square of side 8/9 of the diameter. Denoting the diameter by d, this amounts to the formula $A = [(8/9)d]^2 = (64/81)d^2$. If we compare this to the formula $A = \pi d^2/4$, we find that the Egyptians used the value $\pi = 256/81 = 3.16049$, in error of only 0.6 percent of the true value. A remarkable achievement![13]

✧   ✧   ✧

Of particular interest to us are Problems 56–60. They deal with that most famous of Egyptian monuments, the pyramids, and all use the word *seked* (see fig. 1).[14] What this word means we shall soon find out.

Problem 56 says: "If a pyramid is 250 cubits high and the side of its base 360 cubits long, what is its seked?" Ahmes's solution follows:

Take 1/2 of 360; it makes 180. Multiply 250 so as to get 180; it makes 1/2 1/5 1/50 of a cubit. A cubit is 7 palms. Multiply 7 by 1/2 1/5 1/50:

1	7		
1/2	3	1/2	
1/5	1	1/3	1/15
1/50		1/10	1/25

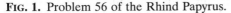

FIG. 1. Problem 56 of the Rhind Papyrus.

The seked is $5\frac{1}{25}$ palms [that is, $(3 + 1/2) + (1 + 1/3 + 1/15) + (1/10 + 1/25) = 5\frac{1}{25}$].[15]

Let us analyze the solution. Clearly 1/2 of 360, or 180, is half the side of the square base of the pyramid (fig. 2). "Multiply 250 so as to get 180" means to find a number x such that 250 times x equals 180. This gives us $x = 180/250 = 18/25$. But Egyptian mathematics required that all answers be given in unit fractions; and the sum of the unit fractions 1/2, 1/5, and 1/50 is indeed 18/25. This number, then, is the ratio of half the side of the base of the pyramid to its height, or the run-to-rise ratio of its face. In effect, the quantity that Ahmes found, the seked, is the *cotangent of the angle between the base of the pyramid and its face*.[16]

Two questions immediately arise: First, why didn't he find the *reciprocal* of this ratio, or the rise-to-run ratio, as we would do today? The answer is that when building a vertical structure, it is natural to measure the *horizontal* deviation from the vertical line for each unit increase in height, that is, the run-to-rise ratio. This indeed is the practice in architecture, where one uses the

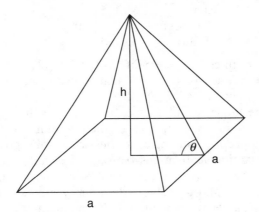

FIG. 2. Square-based pyramid.

term *batter* to measure the inward slope of a supposedly vertical wall.

Second, why did Ahmes go on to multiply his answer by 7? For some reason the pyramid builders measured horizontal distances in "palms" or "hands" and vertical distances in cubits. One cubit equals 7 palms. Thus the required seked, $5\frac{1}{25}$, gives the run-to-rise ratio in units of palms per cubit. Today, of course, we think of these ratios as a pure numbers.

Why was the run-to-rise ratio considered so important as to deserve a special name and four problems devoted to it in the papyrus? The reason is that it was crucial for the pyramid builders to maintain a constant slope of each face relative to the horizon. This may look easy on paper, but once the actual construction began, the builders constantly had to check their progress to ensure that the required slope was maintained. That is, the seked had to be the same for each one of the faces.

Problem 57 is the inverse problem: we are given the seked and the side of a base and are asked to find the height. Problems 58 and 59 are similar to Problem 56 and lead to a seked of $5\frac{1}{4}$ palms (per cubit), except that the answer is given as 5 palms and 1 "finger" (there being 4 fingers in a palm). Finally, Problems 60 asks to find the seked of a pillar 30 cubits high whose base is 15 cubits. We do not know if this pillar had the shape of a pyramid or a cylinder (in which case 15 is the diameter of the base); in either case the answer is 1/4.

The seked found in Problem 56, namely 18/25 (in dimensionless units) corresponds to an angle of 54° 15′ between the base and face. The seked found in Problems 58–59, when converted back to dimensionless units, is $(5\frac{1}{4})$: 7 or 3/4, corresponding to an angle of 53° 8′. It is interesting to compare these figures to the actual angles of some of pyramids at Giza:[17]

Cheops:	51°52′
Chephren:	52°20′
Mycerinus:	50°47′

The figures are in close agreement. As for the pillar in Problem 60, its angle is much larger, as of course we expect of such a structure: $\phi = \cot^{-1}(1/4) = 75° 58′$.

It would be ludicrous, of course, to claim that the Egyptians invented trigonometry. Nowhere in their writings does there appear the concept of an angle, so they were in no position to formulate quantitative relations between the angles and sides of a triangle. And yet (to quote Chase) "at the beginning of the 18th century B.C., and probably a thousand years earlier, when the great pyramids were built, the Egyptian mathematicians had some notion of referring a right triangle to a similar triangle, one of whose sides was a unit of measure, as a standard." We may therefore be justified in crediting the Egyptians with a crude knowledge of practical trigonometry—perhaps "proto-trigonometry" would be a better word—some two thousand years before the Greeks took up this subject and transformed it into a powerful tool of applied mathematics.

NOTES AND SOURCES

1. The papyrus also contains three fragmentary pieces of text unrelated to mathematics, which some authors number as Problems 85, 86, and 87. These are described in Arnold Chase, *The Rhind Mathematical Papyrus: Free Translation and Commentary with Selected Photographs, Transcriptions, Transliterations and Literal Translations* (Reston, VA: National Council of Teachers of Mathematics, 1979), pp. 61–62.

2. *The Rhind Mathematical Papyrus, British Museum 10057 and 10058: Introduction, Transcription, Translation and Commentary* (London, 1923).

3. Chase, *Rhind Mathematical Papyrus*. This extensive work is a reprint, with minor changes, of the same work published by the Mathematical Association of America in two volumes in 1927 and 1929. It contains detailed commentary and an extensive bibliography, as well as numerous color plates of text material. For a biographical sketch of Chase, see the article "Arnold Buffum Chase" in the *American Mathematical Monthly*, vol. 40 (March 1933), pp. 139–142. Other good sources on Egyptian mathematics are Richard J. Gillings, *Mathematics in the Time of the Pharaohs* (1972; rpt. New York: Dover, 1982); George Gheverghese Joseph, *The Crest of the Peacock: Non-European Roots of Mathematics* (Harmondsworth, U.K.: Penguin Books, 1991), chap. 3; Otto Neugebauer, *The Exact Sciences in Antiquity* (1957; rpt. New York: Dover, 1969), chap. 4; and Baertel L. van der Waerden, *Science Awakening*, trans. Arnold Dresden (New York: John Wiley, 1963), chap. 1.

4. Chase, *Rhind Mathematical Papyrus*, p. 27. The royal title "Re" is pronounced "ray."

5. Another important document from roughly the same period is the Golenishchev or Moscow Papyrus, a scroll about the same length as the Rhind Papyrus but only three inches wide. It contains 25 problems and is of poorer quality than the Rhind Papyrus. See Gillings, *Mathematics*, pp. 246–247; Joseph, *Crest of the Peacock*, pp. 84–89; van der Waerden, *Science Awakening*, pp. 33–35; and Carl B. Boyer, *A History of Mathematics* (1968; rev. ed. New York: John Wiley, 1989), pp. 22–24. References to other Egyptian mathematical documents can be found in Chase, *Rhind Mathematical Papyrus*, p. 67; Gillings, *Mathematics*, chaps. 9, 14, and 22; Joseph, *Crest of the Peacock*, pp. 59–61, 66–67 and 78–79; and Neugebauer, *Exact Sciences*, pp. 91–92;

6. As quoted by van der Waerden, *Science Awakening*, p. 16, who apparently quoted from Peet. This differs slightly from Chase's free translation (*Rhind Mathematical Papyrus*, p. 27).

7. Van der Waerden, *Science Awakening*, pp. 16–17.

8. Note that the decomposition is not unique: 7/10 can also be written as $1/5 + 1/2$.

9. For a more detailed discussion of the Egyptians' use of unit fractions, see Boyer, *History of Mathematics*, pp. 15–17; Chase, *Rhind Mathematical Papyrus*, pp. 9–17; Gillings, *Mathematics*, pp. 20–23; and van der Waerden, *Science Awakening*, pp. 19–26.

10. Chase, *Rhind Mathematical Papyrus*, pp. 15–16; van der Waerden, *Science Awakening*, pp. 27–29.

11. See Gillings, *Mathematics*, pp. 154–161.

12. Chase, *Rhind Mathematical Papyrus*, p. 46. For a discussion of Egyptian measures, see ibid., pp. 18–20; Gillings, *Mathematics*, pp. 206–213.

13. The Egyptian value can be conveniently written as $(4/3)^4$. Gillings (*Mathematics*, pp. 139–153) gives a convincing theory as to how Ahmes derived the formula $A = [(8/9)d]^2$ and credits him as being "the first authentic circle-squarer in recorded history!" See also Chase, *Rhind Mathematical Papyrus*, pp. 20–21, and Joseph, *Crest of the Peacock*, pp. 82–84 and 87–89. Interestingly the Babylonians, whose mathematical skills generally exceeded those of the Egyptians, simply equated the area of a circle to the area of the inscribed regular hexagon, leading to $\pi = 3$; see Joseph, *Crest of the Peacock*, p. 113.

14. Pronounced "saykad" or "sayket."

15. Chase, *Rhind Mathematical Papyrus*, p. 51.

16. See, however, ibid., pp. 21–22 for an alternative interpretation.

17. Gillings, *Mathematics*, p. 187.

Recreational Mathematics in Ancient Egypt

roblem 79 of the Rhind Papyrus says (fig. 3):[1]

A house inventory:		houses	7
1	2,801	cats[a]	49
2	5,602	mice	343
4	11,204	spelt	2,301[b]
		hekat	16,807
Total	19,607	Total	19,607

[a]The Egyptian word for "cat" is *myw*; when the missing vowels are inserted, this becomes *meey'auw*.

[b]Obviously Ahmes made a mistake here. The correct entry should be 2,401.

What is the meaning behind this cryptic verse? Clearly we have before us a geometric progression whose initial term and common ratio are both 7, and the scribe shows us how to find its sum. But as any good teacher would do to break the monotony of a routine math class, Ahmes embellishes the exercise with a little story which might be read like this: There are seven houses; in each house there are seven cats; each cat eats seven mice; each mouse eats seven ears of spelt; each ear of spelt produces seven hekat of grain. Find the total number of items involved.

The right hand column clearly gives the terms of the progression $7, 7^2, 7^3, 7^4, 7^5$ followed by their sum, 19,607 (whether the mistaken entry 2,301 was Ahmes's own error in copying or whether it had already been in the original document, we shall never know). But now Ahmes plays his second card: in the left-hand column he shows us how to obtain the answer in a shorter, "clever" way; and in following it we can see the Egyptian method of multiplication at work. The Egyptians knew that any integer can be represented as a sum of terms of the geometric progression $1, 2, 4, 8, \ldots$, and that the representation is unique (this is precisely the representation of an integer in terms of the base 2, the coefficients, or "binary digits," being 0 and 1). To multiply, say, 13 by 17, they only had to write one of the multipliers, say 13, as a sum of powers of 2, $13 = 1 + 4 + 8$,

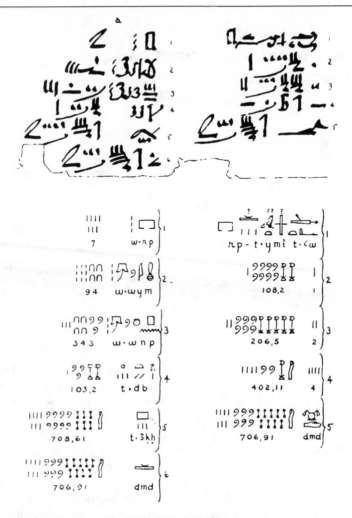

FIG. 3. Problem 79 of the Rhind Papyrus.

multiply each power by the other multiplier, and add the results: $13 \times 17 = 1 \times 17 + 4 \times 17 + 8 \times 17 = 17 + 68 + 136 = 221$. The work can be conveniently done in a tabular form:

$$17 \times 1 = 17 \quad *$$
$$\times 2 = 34$$
$$\times 4 = 68 \quad *$$
$$\times 8 = 136 \quad *$$

The astrisks indicate the powers to be added. Thus the Egyptians could do any multiplication by repeated doubling and adding. In all the Egyptian mathematical writings known to us, this practice is always followed; it was as basic to the Egyptian scribe as the multiplication table is to a pupil today.

So where does 2,801, the first number in the left-hand column of Problem 79, come from? Here Ahmes uses a property of geometric progressions with which the Egyptians were familiar: the sum of the first n terms of a geometric progression with the same initial term and common ratio is equal to the common ratio multiplied by one plus the sum of the first $(n-1)$ terms; in modern notation, $a + a^2 + a^3 + \ldots + a^n = a(1 + a + a^2 + \ldots + a^{n-1})$. This sort of "recursion formula" enabled the Egyptian scribe to reduce the summation of one geometric progression to that of another one with fewer (and smaller) terms. To find the sum of the progression $7 + 49 + 343 + 2,401 + 16,807$, Ahmes thought of it as $7 \times (1 + 7 + 49 + 343 + 2,401)$; since the sum of the terms inside the parentheses is 2,801, all he had to do was to multiply this number by 7, thinking of 7 as $1 + 2 + 4$. This is what the left-hand column shows us. Note that this column requires only three steps, compared to the five steps of the "obvious" solution shown in the right-hand column; clearly the scribe included this exercise as an example in creative thinking.

One may ask: why did Ahmes choose the common ratio 7? In his excellent book, *Mathematics in the Times of the Pharaohs*, Richard J. Gillings answers this question as follows: "The number 7 often presents itself in Egyptian multiplication because, by regular doubling, the first three multipliers are *always* 1, 2, 4, which add to 7."[2] This explanation, however, is somewhat unconvincing, for it would equally apply to 3 ($= 1 + 2$), to 15 ($= 1 + 2 + 4 + 8$), and in fact to all integers of the form $2^n - 1$. A more plausible explanation might be that 7 was chosen because a larger number would have made the calculation too long, while a smaller one would not have illustrated the rapid growth of the progression: had Ahmes used 3, the final answer (363) may not have been "sensational" enough to impress the reader.

The dramatic growth of a geometric progression has fascinated mathematicians throughout the ages; it even found its way into the folklore of some cultures. An old legend has it that the king of Persia was so impressed by the game of chess that he wished to reward its inventor. When summoned to the royal palace, the inventor, a poor peasant from a remote corner of the kingdom, merely requested that one grain of wheat be put on the first square of the chessboard, two grains on the second square, four grains on the third, and so on until all 64 squares were covered. Surprised by the modesty of this request, the king ordered his servants to bring a few bags of wheat, and they patiently began to put the grains on the board.

It soon became clear, however, that not even the entire amount of grain in the kingdom sufficed to fulfill the request, for the sum of the progression $1 + 2 + 2^2 + \ldots + 2^{63}$ is a staggering 18,446,744,073,709,551,615—enough to form a line of grain some two light years long!

Ahmes's Problem 79 has a strong likeness to an old nursery rhyme:

As I was going to St. Ives,
I met a man with seven wives;
Every wife had seven sacks,
Every sack had seven cats,
Every cat had seven kits.
Kits, cats, sacks and wives,
How many were going to St. Ives?

In Leonardo Pisano's ("Fibonacci") famous work *Liber Abaci* (1202) there is a problem which, except for the story involved, is identical to this rhyme. This has led some scholars to suggest that Problem 79 "has perpetuated itself through all the centuries from the times of the ancient Egyptians."[3] To which Gillings replies: "All the available evidence for this [conclusion] is here before us, and one is entitled to draw whatever conclusions one wishes. It is indeed tempting to be able to say to a child, 'Here is a nursery rhyme that is nearly 4,000 years old!' But is it really? We shall never truly know."[4]

Geometric progressions may seem quite removed from trigonometry, but in chapter 9 we will show that the two are indeed closely related. This will allow us to investigate these progressions geometrically and perhaps justify the adjective "geometric" that has, for no apparent reason, been associated with them.

Notes and Sources

1. Arnold Buffum Chase, *The Rhind Mathematical Papyrus: Free Translation and Commentary with Selected Photographs, Transcriptions, Transliterations and Literal Translations* (Reston, Va.: National Council of Teachers of Mathematics, 1979), p. 136. I have used here Chase's literal (rather than free) translation in order to preserve the flavor and charm of the problem as originally stated. This also includes Ahmes's obvious error in the fourth line of the right column. For Chase's free translation, see p. 59 of his book.

2. Richard J. Gillings, *Mathematics in the Times of the Pharoahs.* (1972; rpt. New York: Dover, 1982), p. 168.

3. L. Rodet as quoted by Chase, *Rhind Mathematical Papyrus*, p. 59.

4. Gillings, *Mathematics*, p. 170.

1

Angles

A plane angle is the inclination to one another of two lines in a plane which meet one another and do not lie in a straight line.
—Euclid, *The Elements*, Definition 8

Geometric entities are of two kinds: those of a strictly qualitative nature, such as a point, a line, and a plane, and those that can be assigned a numerical value, a measure. To this last group belong a line segment, whose measure is its length; a planar region, associated with its area; and a rotation, measured by its angle.

There is a certain ambiguity in the concept of angle, for it describes both the qualitative idea of "separation" between two intersecting lines, and the numerical value of this separation—the measure of the angle. (Note that this is not so with the analogous "separation" between two points, where the phrases *line segment* and *length* make the distinction clear.) Fortunately we need not worry about this ambiguity, for trigonometry is concerned only with the quantitative aspects of line segments and angles.[1]

The common unit of angular measure, the degree, is believed to have originated with the Babylonians. It is generally assumed that their division of a circle into 360 parts was based on the closeness of this number to the length of the year, 365 days. Another reason may have been the fact that a circle divides naturally into six equal parts, each subtending a chord equal to the radius (fig. 4). There is, however, no conclusive evidence to support these hypotheses, and the exact origin of the 360-degree system may remain forever unknown.[2] In any case, the system fitted well with the Babylonian sexagesimal (base 60) numeration system, which was later adopted by the Greeks and used by Ptolemy in his table of chords (see chapter 2).

As a numeration system, the sexagesimal system is now obsolete, but the division of a circle into 360 parts has survived—

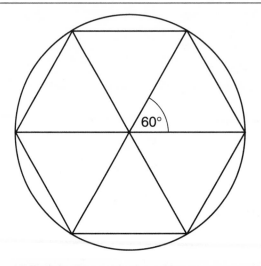

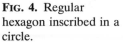

FIG. 4. Regular hexagon inscribed in a circle.

not only in angular measure but also in the division of an hour into 60 minutes and a minute into 60 seconds. This practice is so deeply rooted in our daily life that not even the ascendancy of the metric system was able to dispel it, and Florian Cajori's statement in *A History of Mathematics* (1893) is still true today: "No decimal division of angles is at the present time threatened with adoption, not even in France [where the metric system originated]."[3] Nevertheless, many hand-held calculators have a GRAD option in which a right angle equals 100 "gradians," and fractional parts of a gradian are reckoned decimally.

The word *degree* originated with the Greeks. According to the historian of mathematics David Eugene Smith, they used the word μοιρα (moira), which the Arabs translated into *daraja* (akin to the Hebrew *dar'ggah*, a step on a ladder or scale); this in turn became the Latin *de gradus*, from which came the word degree. The Greeks called the sixtieth part of a degree the "first part," the sixtieth part of that the "second part," and so on. In Latin the former was called *pars minuta prima* ("first small part") and the latter *pars minuta secunda* ("second small part"), from which came our *minute* and *second*.[4]

In more recent times the *radian* or *circular measure* has been universally adopted as the natural unit of angular measure. One radian is the angle, measured at the center of a circle, that subtends an arc length of one radius along the circumference (fig. 5). Since a complete circle encompasses $2\pi(\approx 6.28)$ radii along the circumference, and each of these radii corresponds to a central angle of 1 radian, we have $360° = 2\pi$ radians; hence 1 radian $= 360°/2\pi \approx 57.29°$. The oft-heard statement that a radian is a more convenient unit than a degree because it is

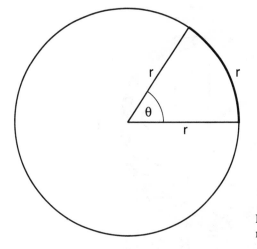

FIG. 5. Radian angular measure.

larger, and thus allows us to express angles by smaller numbers, is simply not true.[5] The sole reason for using radians is that it simplifies many formulas. For example, a circular arc of angular width θ (where θ is in radians) subtends an arc length given by $s = r\theta$; but if θ is in degrees, the corresponding formula is $s = \pi r\theta/180$. Similarly, the area of a circular sector of angular width θ is $A = r^2\theta/2$ for θ in radians and $A = \pi r^2\theta/360$ for θ in degrees.[6] The use of radians rids these formulas of the "unwanted" factor $\pi/180$.

Even more important, the fact that a small angle and its sine are nearly equal numerically—the smaller the angle, the better the approximation—holds true only if the angle is measured in radians. For example, using a calculator we find that the sine of one degree ($\sin 1°$) is 0.0174524; but if the $1°$ is converted to radians, we have $1° = 2\pi/360° \approx 0.0174533$, so the angle and its sine agree to within one hundred thousandth. For an angle of $0.5°$ (again expressed in radians) the agreement is within one millionth, and so on. It is this fact, expressed as $\lim_{\theta\to 0}(\sin\theta)/\theta = 1$, that makes the radian measure so important in calculus.

The word *radian* is of modern vintage; it was coined in 1871 by James Thomson, brother of the famous physicist Lord Kelvin (William Thomson); it first appeared in print in examination questions set by him at Queen's College in Belfast in 1873.[7] Earlier suggestions were "rad" and "radial."

No one knows where the convention of measuring angles in a counterclockwise sense came from. It may have originated with our familiar coordinate system: a $90°$ counterclockwise turn takes us from the positive x-axis to the positive y-axis, but the

Chords

When considered separately, line segments and angles behave in a simple manner: the combined length of two line segments placed end-to-end along the same line is the sum of the individual lengths, and the combined angular measure of two rotations about the same point in the plane is the sum of the individual rotations. It is only when we try to relate the two concepts that complications arise: the equally spaced rungs of a ladder, when viewed from a fixed point, do not form equal angles at the observer's eye (fig. 7), and conversely, equal angles, when projected onto a straight line, do not intercept equal segments (fig. 8). Elementary plane trigonometry—roughly speaking, the trigonometry known by the sixteenth century—concerns itself with the quantitative relations between angles and line segments, particularly in a triangle; indeed, the very word "trigonometry" comes from the Greek words *trigonon* = triangle, and *metron* = measure.[1]

As we have seen, the Egyptians used a kind of primitive trigonometry as early as the second millennium B.C. in building their pyramids. In Mesopotamia, Babylonian astronomers kept meticulous records of the rising and setting of stars, of the motion of the planets and of solar and lunar eclipses, all of which required familiarity with angular distances measured on the celestial sphere.[2] The *gnomon*, a simple device for telling the hour from the length of the shadow cast by a vertical rod, was known to the early Greeks, who, according to the historian Herodotus (ca. 450 B.C.), got it from the Babylonians. The gnomon is essentially an analog device for computing the cotangent function:

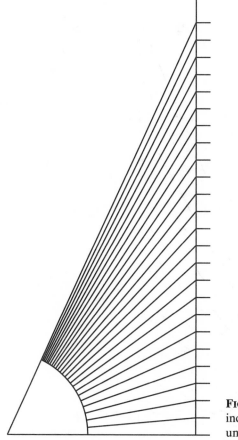

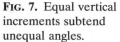

FIG. 7. Equal vertical increments subtend unequal angles.

if (see fig. 9) h denotes the height of the rod and s the length of its shadow when the sun is at an altitude of α degrees above the horizon, then $s = h\cot\alpha$, so that s is proportional to $\cot\alpha$. Of course, the ancients were not interested in the cotangent function as such but rather in using the device as a timekeeper; in fact, by measuring the daily variation of the shadow's length at noon, the gnomon could also be used to determine the day of the year.

Thales of Miletus (ca. 640–546 B.C.), the first of the long line of Greek philosophers and mathematicians, is said to have measured the height of a pyramid by comparing the shadow it casts with that of a gnomon. As told by Plutarch in his *Banquet of the Seven Wise Men*, one of the guests said to Thales:

Whereas he [the king of Egypt] honors you, he particularly admires you for the invention whereby, with little effort and by the aid of no

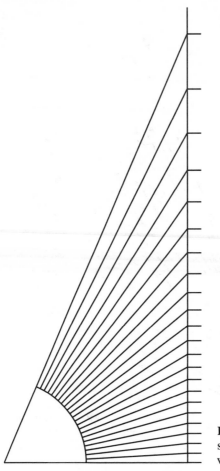

FIG. 8. Equal angles subtend unequal vertical increments.

mathematical instrument, you found so accurately the height of the pyramids. For, having fixed your staff erect at the point of the shadow cast by the pyramid, two triangles were formed by the tangent rays of the sun, and from this you showed that the ratio of one shadow to the other was equal to the ratio of the [height of the] pyramid to the staff.[3]

Again trigonometry was not directly involved, only the similarity of two right triangles. Still, this sort of "shadow reckoning" was fairly well known to the ancients and may be said to be the precursor of trigonometry proper. Later, such simple methods were successfully applied to measure the dimensions of the earth, and later still, the distance to the stars (see chapter 5).

Trigonometry in the modern sense of the word began with Hipparchus of Nicaea (ca.190–120 B.C.), considered the great-

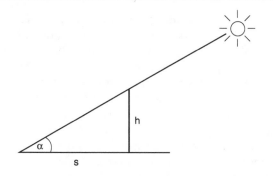

FIG. 9. Gnomon.

est astronomer of antiquity. As with many of the Greek scholars, Hipparchus's work is known to us mainly through references by later writers, in this case the commentary on Ptolemy's *Almagest* by Theon of Alexandria (ca. 390 A.D.). He was born in the town of Nicaea (now Iznik in northwest Turkey) but spent most of his life on the island of Rhodes in the Aegean Sea, where he set up an observatory. Using instruments of his own invention, he determined the positions of some 1,000 stars in terms of their celestial longitude and latitude and recorded them on a map— the first accurate star atlas (he may have been led to this project by his observation, in the year 134 B.C. of a nova—an explod- ing star that became visible where none had been seen before). To classify stars according to their brightness, Hipparchus in- troduced a scale in which the brightest stars were given magni- tude 1 and the faintest magnitude 6; this scale, though revised and greatly extended in range, is still being used today. Hip- parchus is also credited with discovering the precession of the equinoxes—a slow circular motion of the celestial poles once every 26,700 years; this apparent motion is now known to be caused by a wobble of the earth's own axis (it was Newton who correctly explained this phenomenon on the basis of his theory of gravitation). And he refined and simplified the old system of epicylces, invented by Aristotle to explain the observed motion of the planets around the earth (see chapter 7); this was actu- ally a retreat from his predecessor Aristarchus, who had already envisioned a universe in which the sun, and not the earth, was at the center.

To be able to do his calculations Hipparchus needed a table of trigonometric ratios, but he had nowhere to turn: no such ta- ble existed, so he had to compute one himself. He considered every triangle—planar or spherical—as being inscribed in a cir- cle, so that each side becomes a chord. In order to compute the various parts of the triangle one needs to find the length of the chord as a function of the central angle, and this became the chief task of trigonometry for the next several centuries. As an astronomer, Hipparchus was chiefly concerned with spherical triangles, but he must have known many of the formulas of plane trigonometry, among them the identities (in modern notation) $\sin^2 \alpha + \cos^2 \alpha = 1$, $\sin^2 \alpha/2 = (1 - \cos \alpha)/2$, and $\sin(\alpha \pm \beta) = \sin \alpha \cos \beta \pm \cos \alpha \sin \beta$. These formulas, of course, were derived by purely geometric means and expressed as theorems about the angles and chords in a circle (the first formula, for example, is the trigonometric equivalent of the Pythagorean Theorem); we will return to some of these formulas in chapter 6. Hipparchus wrote twelve books on the computation of chords in a circle, but all are lost.

The first major work on trigonometry to have come to us intact is the *Almagest* by Claudius Ptolemaeus, commonly known as Ptolemy (ca. 85–ca. 165 A.D.).[4] Ptolemy lived in Alexandria, the intellectual center of the Hellenistic world, but details of his life are lacking (he is unrelated to the Ptolemy dynasty that ruled Egypt after the death of Alexander the Great in 323 B.C.). In contrast to most of the Greek mathematicians, who regarded their discipline as a pure, abstract science, Ptolemy was first and foremost an applied mathematician. He wrote on astronomy, geography, music, and possibly also optics. He compiled a star catalog based on Hipparchus's work, in which he listed and named forty-eight constellations; these names are still in use today. In his work *Geography*, Ptolemy systematically used the technique of map projection (a system for mapping the spherical earth onto a flat sheet of paper), which Hipparchus had already introduced; his map of the then known world, complete with a grid of longitude and latitude, was the standard world map well into the Middle Ages (fig. 10). However, Ptolemy seriously underestimated the size of the earth, rejecting Eratosthenes' correct estimate as being too large (see chapter 5). In hindsight this turned out to be a blessing, for it spurred Columbus to attempt a westward sea voyage from Europe to Asia, an endeavor which brought about the discovery of the New World.

Ptolemy's greatest work is the *Almagest*, a summary of mathematical astronomy as it was known in his time, based on the assumption of a motionless earth seated at the center of the universe and the heavenly bodies moving around it in their prescribed orbits (the geocentric system). The *Almagest* consists of thirteen parts ("books") and is thus reminiscent of the thirteen books of Euclid's *Elements*. The similarity goes even further, for the two works contain few of their authors' own discover-

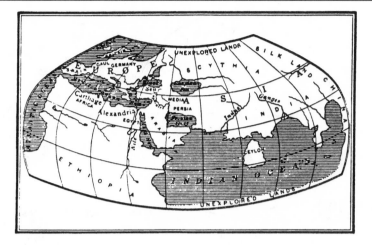

FIG. 10. Ptolemy's world map.

ies; rather, they are compilations of the state of knowledge of their respective fields and are thus based on the achievements of their predecessors (in Ptolemy's case, mainly Hipparchus). Both works have exercised an enormous influence on generations of thinkers; but unlike the *Elements*, which to this day forms the core of classical geometry, the *Almagest* lost much of its authority once Copernicus's heliocentric system was accepted. As a consequence, it is much less known today than the *Elements*— an unfortunate state of affairs, for the *Almagest* is a model of exposition that could well serve as an example even to modern writers.

The word *Almagest* had an interesting evolution: Ptolemy's own title, in translation, was "mathematical syntaxis," to which later generations added the superlative *megiste* ("greatest"). When the Arabs translated the work into their own language, they kept the word *megiste* but added the conjunction *al* ("the"), and in due time it became known as the *Almagest*.[5] In 1175 the Arab version was translated into Latin, and from then on it became the cornerstone of the geocentric world picture; it would dominate the scientific and philosophical thinking of Europe well into the sixteenth century and would become the canon of the Roman Church.

✧   ✧   ✧

Of special interest to us here is Ptolemy's table of chords, which is the subject of chapters 10 and 11 of the first book of the *Almagest*. His table gives the length of a chord in a circle

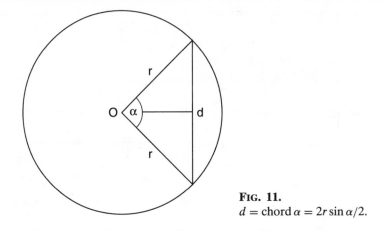

FIG. 11.
$d = \text{chord } \alpha = 2r \sin \alpha/2.$

as a function of the central angle that subtends it (fig. 11) for angles from 0° to 180° at intervals of half a degree. A moment's thought will show that this is essentially a table of sines: denoting the radius by r, the central angle by α, and the length of the chord by d, we have

$$d = 2r \sin \frac{\alpha}{2}. \tag{1}$$

Ptolemy takes the diameter of the circle to be 120 units, so that $r = 60$ (the reason for this choice will soon become clear). Equation (1) then becomes $d = 120 \sin \alpha/2$. Thus, apart from the proportionality factor 120, we have a table of values of $\sin \alpha/2$ and therefore (by doubling the angle) of $\sin \alpha$.

In computing his table Ptolemy used the Babylonian sexagesimal or base 60 numeration system, the only suitable system available in his day for handling fractions (the decimal system was still a thousand years in the future). But he used it in conjunction with the Greek system in which each letter of the alphabet is assigned a numerical value: $\alpha = 1$, $\beta = 2$, and so on. This makes the reading of his table a bit cumbersome, but with a little practice one can easily become proficient at it (fig. 12). For example, for an angle of 7° (expressed by the Greek letter ζ), Ptolemy's table gives a chord length of 7; 19, 33 (written as ζ ιθ λγ, the letters ι, θ, λ, and γ representing 10, 9, 30, and 3, respectively), which is the modern notation for the sexagesimal number $7 + 19/60 + 33/3{,}600$ (the semicolon is used to separate the integral part of the number from its fractional part, and the commas separate the sexagesimal positions). When written in our decimal systems, this number is very nearly equal to 7.32583; the true length of the chord, rounded to five places, is 7.32582. Quite a remarkable achievement!

Κανόνιον τῶν ἐν κύκλῳ εὐθειῶν			Table of Chords		
περιφε-ρειῶν	εὐθειῶν	ἐξηκοστῶν	arcs	chords	sixtieths
∠'	σ λα κε	σ α β ν	½°	0;31,25	0;1,2,50
α	α β ν	σ α β ν	1°	1;2,50	0;1,2,50
α∠'	α λδ ιε	σ α β ν	1½°	1;34,15	0;1,2,50
β	β ε μ	σ α β ν	2°	2;5,40	0;1,2,50
β∠'	β λζ δ	σ α β μη	2½°	2;37,4	0;1,2,48
γ	γ η κη	σ α β μη	3°	3;8,28	0;1,2,48
γ∠'	γ λθ νβ	σ α β μη	3½°	3;39,52	0;1,2,48
δ	δ ια ις	σ α β μζ	4°	4;11,16	0;1,2,47
δ∠'	δ μβ μ	σ α β μζ	4½°	4;42,40	0;1,2,47
ε	ε ιδ δ	σ α β μς	5°	5;14,4	0;1,2,46
ε∠'	ε με κζ	σ α β με	5½°	5;45,27	0;1,2,45
ϛ	ϛ ις μθ	σ α β μδ	6°	6;16,49	0;1,2,44
ϛ∠'	ϛ μη ια	σ α β μγ	6½°	6;48,11	0;1,2,43
ζ	ζ ιθ λγ	σ α β μβ	7°	7;19,33	0;1,2,42
ζ∠'	ζ ν νδ	σ α β μα	7½°	7;50,54	0;1,2,41
⋮	⋮	⋮	⋮	⋮	⋮
ροδ∠'	ριθ να μγ	σ σ β νγ	174½°	119;51,43	0;0,2,53
ροε	ριθ νγ ι	σ σ β λς	175°	119;53,10	0;0,2,36
ροε∠'	ριθ νδ κζ	σ σ β κ	175½°	119;54,27	0;0,2,20
ροϛ	ριθ νε λη	σ σ β γ	176°	119;55,38	0;0,2,3
ροϛ∠'	ριθ νς λθ	σ σ α μζ	176½°	119;56,39	0;0,1,47
ροζ	ριθ νζ λβ	σ σ α λ	177°	119;57,32	0;0,1,30
ροζ∠'	ριθ νη ιη	σ σ α ιδ	177½°	119;58,18	0;0,1,14
ροη	ριθ νη νε	σ σ σ νζ	178°	119;58,55	0;0,0,57
ροη∠'	ριθ νθ κδ	σ σ σ μα	178½°	119;59,24	0;0,0,41
ροθ	ριθ νθ μδ	σ σ σ κε	179°	119;59,44	0;0,0,25
ροθ∠'	ριθ νθ νς	σ σ σ θ	179½°	119;59,56	0;0,0,9
ρπ	ρκ σ σ	σ σ σ σ	180°	120;0,0	0;0,0,0

FIG. 12. A section from Ptolemy's table of chords.

Ptolemy's table gives the chord length to an accuracy of two sexagesimal places, or 1/3,600, which is sufficient for most applications even today. Moreover, the table has a column of "sixties" that allows one to interpolate between successive entries: it gives the mean increment in the chord length from one entry to the next, that is, the increment divided by 30 (the interval between successive angles, measured in minutes of arc).[6] In computing his table, Ptolemy used the formulas mentioned earlier in connection with Hipparchus, all of which are proved in the *Almagest*.[7]

Ptolemy now shows how the table can be used to solve any planar triangle, provided at least one side is known. Following Hipparchus, he considers the triangle to be inscribed in a circle. We will show here the simplest case, that of a right triangle.[8] Let

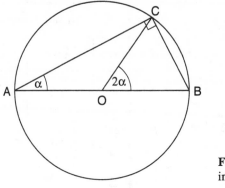

FIG. 13. Right triangle inscribed in a circle.

the triangle be *ABC* (fig. 13), with the right angle at *C*. From elementary geometry we know that the hypotenuse $c = AB$ is the diameter of the circle passing through *A*, *B*, and *C*. Denoting by *O* the center of this circle (that is, the midpoint of *AB*), a well-known theorem says that $\angle BOC = 2\angle BAC = 2\alpha$. Suppose that α and *c* are given. We first compute 2α and use the table to find the length of the corresponding chord; and since the table assumes $c = 120$, we have to multiply the length by the ratio $c/120$. This gives us the side $a = BC$. The remaining side $b = AC$ can then be found from the Pythagorean Theorem, and the angle $\beta = \angle ABC$ from the equation $\beta = 90° - \alpha$. Conversely, if two sides are known, say *a* and *c*, we compute the ratio a/c, multiply it by 120, and then use the table in reverse to find 2α and thence α.

The procedure can be summarized in the formula

$$a = \frac{c}{120} \text{ chord } 2\alpha, \tag{2}$$

where chord 2α is the length of the chord whose central angle is 2α. This leads to an interesting observation: in the sexagesimal (base 60) system, multiplying and dividing by 120 is analogous to multiplying and dividing by 20 in the *decimal* system: we simply multiply or divide by 2 and shift the point one place to the right or left, respectively. Thus equation (2) requires us to double the angle, look up the corresponding chord, and divide it by 2. To do this again and again becomes a chore, so it was only a matter of time before someone shortened this labor by tabulating *half* the chord as a function of *twice* the angle, in other words our modern sine function.[9] This task befell the Hindus.

NOTES AND SOURCES

1. As proof that the relation between angles and line segments is far from simple, consider the following theorem: If two angle-bisectors in a triangle are equal in length, the triangle is isosceles. Looking deceptively simple, its proof can elude even experienced practitioners. See H.S.M. Coxeter, *Introduction to Geometry* (New York: John Wiley, 1969), pp. 9 and 420.

2. For a good summary of Babylonian astronomy, see Otto Neugebauer, *The Exact Sciences in Antiquity* (1957; 2d ed., New York: Dover, 1969), chapter 5.

3. As quoted in David Eugene Smith, *History of Mathematics* (1925; rpt. New York: Dover, 1958), vol. II, pp. 602–603.

4. Asger Aaboe, in *Episodes from the Early History of Mathematics* New York: Random House, 1964), gives his name as Klaudios Ptolemaios, which is closer to the Greek pronunciation. I have used the more common Latin spelling Ptolemaeus.

5. Smith (*History of Mathematics*, vol. I, p. 131) comments that since the prefix "al" means "the," "to speak of 'the *Almagest*' is like speaking of 'the the-greatest.' Nevertheless, the misnomer is so common that I have kept it here.

6. This column is akin to the "proportional parts" column found in a table of logarithms.

7. For a full discussion of how Ptolemy compiled his table, see Aaboe, *Episodes*, pp. 112–126.

8. The other cases can be handled by dissecting the triangle into right triangles; see ibid., pp. 107–111.

9. This becomes clear from the right triangle ABC (fig. 13), in which $a = c \sin \alpha$. Comparing this with equation (2), we have $\sin \alpha = (\text{chord } 2\alpha)/120$.

Plimpton 322: The Earliest Trigonometric Table?

Whereas the Egyptians wrote their records on papyrus and wood and the Chinese on bark and bamboo—all perishable materials—the Babylonians used clay tablets, a virtually indestructible medium. As a result, we are in possession of a far greater number of Babylonian texts than those of any other ancient civilization, and our knowledge of their history—their military campaigns, commercial transactions, and scientific achievements—is that much richer.

Among the estimated 500,000 tablets that have reached museums around the world, some 300 deal with mathematical issues. These are of two kinds: "table texts" and "problem texts," the latter dealing with a variety of algebraic and geometric problems. The "table texts" include multiplication tables and tables of reciprocals, compound interest, and various number sequences; they prove that the Babylonians possessed a remarkably high degree of computational skills.

One of the most intriguing tablets to reach us is known as Plimpton 322, so named because it is number 322 in the G. A. Plimpton Collection at Columbia University in New York (fig. 14). It dates from the Old Babylonian period of the Hammurabi dynasty, roughly 1800–1600 B.C. A careful analysis of the text reveals that it deals with *Pythagorean triples*—integers a, b, c such that $c^2 = a^2 + b^2$; examples of such triples are (3, 4, 5), (5, 12, 13) and (16, 63, 65). Because of the Pythagorean Theorem—or more precisely, its converse—such triples can be used to form right triangles with integer sides.

Unfortunately, the left end of the tablet is damaged and partially missing, but traces of modern glue found at the edge prove that the missing part broke off after the tablet was discovered, and one day it may yet show up in the antiquarian market. Thanks to meticulous scholarly research, the missing part has been partly reconstructed, and we can now read the table with relative ease. We should remember, however, that the Babylonians used the sexagesimal (base 60) numeration system, and that they did not have a symbol for zero; consequently, numbers may be interpreted in different ways, and the correct place value of the individual "digits" must be deduced from the context.

FIG. 14. Plimpton 322.

The text is written in cuneiform (wedged-shaped) characters, which were carved into a wet clay tablet by means of a stylus. The tablet was then baked in an oven or dried in the sun until it hardened to form a permanent record. Table 1 reproduces the text in modern notation, in which sexagesimal "digits" (themselves expressed in ordinary decimal notation) are separated by commas. There are four columns, of which the rightmost, headed by the words "its name" in the original text, merely gives the sequential number of the lines from 1 to 15. The second and third columns (counting from right to left) are headed "solving number of the diagonal" and "solving number of the width," respectively; that is, they give the length of the diagonal and the short side of a rectangle, or equivalently, the length of the hypotenuse and one side in a right triangle. We will label these columns with the letters c and b, respectively. As an example, the first line shows the entries $b = 1, 59$ and $c = 2, 49$, which represent the numbers $1 \times 60 + 59 = 119$ and $2 \times 60 + 49 = 169$. A quick calculation then gives the other side of the triangle as $a = \sqrt{(169^2 - 119^2)} = 120$; hence the triple (119, 120, 169) is a Pythagorean triple. Again, in the third line we read $b = 1, 16, 41 = 1 \times 60^2 + 16 \times 60 + 41 = 4601$ and $c = 1, 50, 49 =$

TABLE 1.

$(c/a)^2$	b	c	
[1,59,0,]15	1,59	2,49	1
[1,56,56,]58,14,50,6,15	56,7	3,12,1	2
[1,55,7,]41,15,33,45	1,16,41	1,50,49	3
[1,]5[3,1]0,29,32,52,16	3,31,49	5,9,1	4
[1,]48,54,1,40	1,5	1,37	5
[1,]47,6,41,40	5,19	8,1	6
[1,]43,11,56,28,26,40	38,11	59,1	7
[1,]41,33,59,3,45	13,19	20,49	8
[1,]38,33,36,36	9,1	12,49	9
1,35,10,2,28,27,24,26,40	1,22,41	2,16,1	10
1,33,45	45	1,15	11
1,29,21,54,2,15	27,59	48,49	12
[1,]27,0,3,45	7,12,1	4,49	13
1,25,48,51,35,6,40	29,31	53,49	14
[1,]23,13,46,40	56	53	15

Note: The numbers in brackets are reconstructed.

$1 \times 60^2 + 50 \times 60 + 49 = 6649$; therefore $a = \sqrt{(6649^2 - 4601^2)} = 4800$, giving the triple (4601, 4800, 6649).

The table contains some obvious errors. In line 9 we find $b = 9,1 = 9 \times 60 + 1 = 541$ and $c = 12,49 = 12 \times 60 + 49 = 769$, and these do not form a Pythagorean triple (the third number a not being an integer). But if we replace the $9,1$ by $8,1 = 481$, we do indeed get the triple (481, 600, 769). It seems that this error was simply a "typo": the scribe must have been momentarily distracted and carved nine marks into his soft clay instead of eight; and once dried in the sun, his oversight became part of recorded history. Again, in line 13 we have $b = 7,12,1 = 7 \times 60^2 + 12 \times 60 + 1 = 25921$ and $c = 4,49 = 4 \times 60 + 49 = 289$, and these do not form a Pythagorean triple; but we may notice that 25921 is the square of 161, and the numbers 161 and 289 do form the triple (161, 240, 289). It seems that the scribe simply forgot to take the square root of 25921. And in row 15 we find $c = 53$, whereas the correct entry should be twice that number, or $106 = 1, 46$, producing the triple (56, 90, 106).[1] These errors leave one with a sense that human nature has not changed over the past 4000 years: our anonymous scribe was no more guilty of negligence than a student begging his or her professor to ignore "just a little stupid mistake" on the exam.[2]

The leftmost column is the most intriguing of all. Its heading again mentions the word "diagonal," but the exact meaning of the remaining text is not entirely clear. However, upon examining its entries a startling fact comes to light: this column gives the square of the ratio (c/a), that is, the values of $\csc^2\alpha$, where α is the angle opposite of side a. Let us verify this for line 1. We have $b = 1,59 = 119$ and $c = 2,49 = 169$, from which we find $a = 120$. Hence $(c/a)^2 = (169/120)^2 = 1.983$, rounded to three decimal places. The corresponding entry in column 4 is $1,59,0,15 = 1 + 59 \times (1/60) + 0 \times (1/60^2) + 15 \times (1/60^3) = 1.983$. (We should note again that the Babylonians did not use a symbol for the "empty slot"—our zero—and therefore a number could be interpreted in many different ways; the correct interpretation must be deduced from the context. In the example just given, we assume that the leading 1 stands for units rather than sixties.) The reader may check other entries in this column and confirm that they are equal to $(c/a)^2$.

Several questions arise: Is the order of entries in the table random, or does it follow some hidden pattern? How did the Babylonians find those particular numbers that form Pythagorean triples? And why were they interested in these numbers—specifically, in the ratio $(c/a)^2$—in the first place? The first question is relatively easy to answer: if one compares the values of $(c/a)^2$ line by line, one discovers that they decrease steadily from 1.983 to 1.387, so it seems likely that the order of entries was determined by this sequence. Moreover, if we compute the square root of each entry in column 4—that is, the ratio $c/a = \csc\alpha$—and then find the corresponding angle α, we discover that α increases steadily from just above 45° to 58°. It thus seems that the author of our text was not only interested in finding Pythagorean triples, but also in determining the ratio c/a of the corresponding right triangles. This hypothesis may one day be confirmed if the missing part of the tablet will show up, as it may well contain the missing columns for a and c/a.

As to how the Pythagorean triples were found, there is only one plausible explanation: the Babylonians must have known the algorithm for generating these triples. Let u and v be any two positive integers such that $u > v$; then the three numbers

$$a = 2uv, b = u^2 - v^2, c = u^2 + v^2 \tag{1}$$

form a Pythagorean triple. (If in addition we require that u and v are of odd parity—one even and the other odd—and that they do not have any common factor, then (a, b, c) is a *primitive* Pythagorean triple, i.e., $a, b,$ and c have no common factor.) It is easy to confirm that the numbers a, b and c as given by equations (1) satisfy the equation $c^2 = a^2 + b^2$; the converse of this statement—that *every* Pythagorean triple can be found in this way—is proved in a standard course in number theory. Plimpton 322 thus shows that the Babylonians were not only familiar with the Pythagorean Theorem a thousand years before Pythagoras, but that they knew the rudiments of number theory and had the computational skills to put the theory into practice.[3]

NOTES AND SOURCES

(The material in this section is based on Otto Neugebauer, *The Exact Sciences in Antiquity* [1957; rpt. New York: Dover, 1969], chap. 2. See also Howard Eves, *An Introduction to the History of Mathematics* [Fort Worth: Saunders College Publishing, 1992], pp. 44–47.)

1. This, however, is not a *primitive triple*, since it can be reduced to the simpler triple (28, 45, 53); the two triples represent similar triangles.

2. A fourth error occurs in line 2, where the entry 3,12,1 should be 1,20,25, producing the triple (3367, 3456, 4825). This error has remained unexplained.

3. As to how the Babylonians did the computations, see Neugebauer, *Exact Sciences*, pp. 39–42.

3

Six Functions Come of Age

It is quite difficult to describe with certainty the
beginning of trigonometry. . . . In general, one may say
that the emphasis was placed first on astronomy, then
shifted to spherical trigonometry, and finally moved on
to plane trigonometry.
—Barnabas Hughes, Introduction to Regiomontanus'
On Triangles

An early Hindu work on astronomy, the *Surya Siddhanta* (ca. 400 A.D.), gives a table of half-chords based on Ptolemy's table (fig. 15). But the first work to refer explicitly to the sine as a function of an angle is the *Aryabhatiya* of Aryabhata (ca. 510), considered the earliest Hindu treatise on pure mathematics.[1] In this work Aryabhata (also known as Aryabhata the elder; born 475 or 476, died ca. 550)[2] uses the word *ardha-jya* for the half-chord, which he sometimes turns around to *jya-ardha* ("chord-half"); in due time he shortens it to *jya* or *jiva*.

Now begins an interesting etymological evolution that would finally lead to our modern word "sine." When the Arabs translated the *Aryabhatiya* into their own language, they retained the word *jiva* without translating its meaning. In Arabic—as also in Hebrew—words consist mostly of consonants, the pronunciation of the missing vowels being understood through common usage. Thus *jiva* could also be pronounced as *jiba* or *jaib*, and *jaib* in Arabic means bosom, fold, or bay. When the Arabic version was translated into Latin, *jaib* was translated into *sinus*, which means bosom, bay, or curve (on lunar maps regions resembling bays are still described as sinus). We find the word *sinus* in the writings of Gherardo of Cremona (ca. 1114–1187), who translated many of the old Greek works, including the *Almagest*, from Arabic into Latin. Other writers followed, and soon the word *sinus*—or *sine* in its English version—became common in mathematical texts throughout Europe. The abbreviated notation *sin* was first used

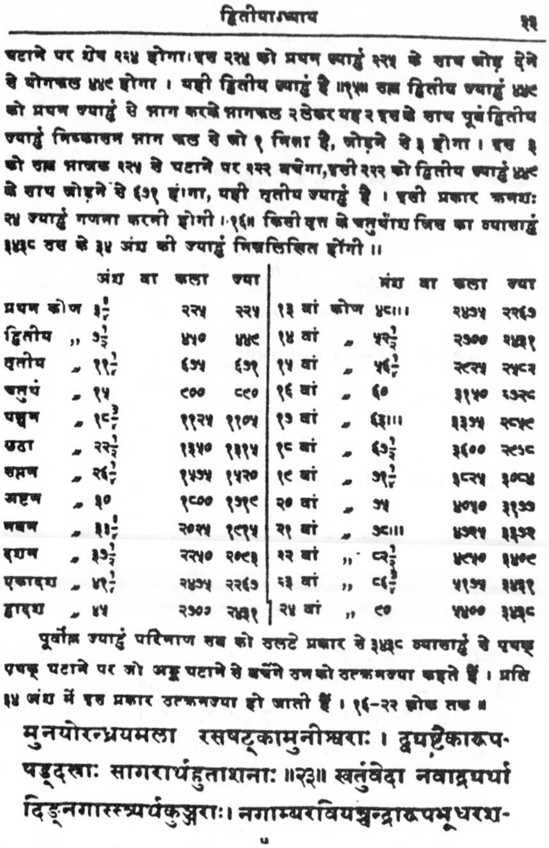

FIG. 15. A page from the Surya Siddhanta.

by Edmund Gunter (1581–1626), an English minister who later became professor of astronomy at Gresham College in London. In 1624 he invented a mechanical device, the "Gunter scale," for computing with logarithms—a forerunner of the familiar slide rule—and the notation *sin* (as well as *tan*) first appeared in a drawing describing his invention.[3]

Mathematical notation often takes unexpected turns. Just as Leibniz objected to William Ougthred's use of the symbol "×" for multiplication (on account of its similarity to the letter x), so did Carl Friedrich Gauss (1777–1855) object to the notation $sin^2\phi$ for the square of sin ϕ:

$sin^2\phi$ is odious to me, even though Laplace made use of it; should it be feared that $sin^2\phi$ might become ambiguous, which would perhaps never occur . . . well then, let us write $(sin\,\phi)^2$, but not $sin^2\phi$, which by analogy should signify sin $(sin\,\phi)$."[4]

Notwithstanding Gauss's objection, the notation $\sin^2 \phi$ has survived, but his concern for confusing it with $\sin(\sin \phi)$ was not without reason: today the repeated application of a function to different initial values is the subject of active research, and expressions like $\sin(\sin(\sin \ldots (\sin \phi) \ldots))$ appear routinely in the mathematical literature.

The remaining five trigonometric functions have a more recent history. The cosine function, which we regard today as equal in importance to the sine, first arose from the need to compute the sine of the complementary angle. Aryabhata called it *kotijya* and used it in much the same way as trigonometric tables of modern vintage did (until the hand-held calculator made them obsolete), by tabulating in the same column the sines of angles from 0° to 45° and the cosines of the complementary angles. The name *cosinus* originated with Edmund Gunter: he wrote *co.sinus*, which was modified to *cosinus* by John Newton (1622–1678), a teacher and author of mathematics textbooks (he is unrelated to Isaac Newton) in 1658. The abbreviated notation *Cos* was first used in 1674 by Sir Jonas Moore (1617–1679), an English mathematician and surveyor.

The functions secant and cosecant came into being even later. They were first mentioned, without special names, in the works of the Arab scholar Abul-Wefa (940–998), who was also one of the first to construct a table of tangents; but they were of little use until navigational tables were computed in the fifteenth century. The first printed table of secants appeared in the work *Canon doctrinae triangulorum* (Leipzig, 1551) by Georg Joachim Rhæticus (1514–1576), who studied with Copernicus and became his first disciple; in this work all six trigonometric functions appear for the first time. The notation *sec* was suggested in 1626 by the French-born mathematician Albert Girard (1595–1632), who spent most of his life in Holland. (Girard was the first to understand the meaning of negative roots in geometric problems; he also guessed that a polynomial has as many roots as its degree, and was an early advocate of the use of parentheses in algebra.) For sec A he wrote $\overset{sec}{A}$, with a similar notation for tan A, but for sin A and cos A he wrote A and a, respectively.

The tangent and cotangent ratios, as we have seen, originated with the gnomon and shadow reckoning. But the treatment of these ratios as functions of an angle began with the Arabs. The first table of tangents and cotangents was constructed around 860 by Ahmed ibn Abdallah al-Mervazi, commonly known as Habash al-Hasib ("the computer"), who wrote on astronomy and astronomical instruments.[5] The astronomer al-Battani (known in Europe as Albategnius; born in Battan,

Mesopotamia, ca. 858, died 929) gave a rule for finding the elevation of the sun above the horizon in terms of the length s of the shadow cast by a vertical gnomon of height h; his rule (ca. 920),

$$s = \frac{h \sin (90° - \phi)}{\sin \phi},$$

is equivalent to the formula $s = h \cot \phi$. Note that in expressing it he used only the sine function, the other functions having not yet been known by name. (It was through al-Battani's work that the Hindu half-chord function—our modern sine—became known in Europe.) Based on this rule, he constructed a "table of shadows"—essentially a table of cotangents—for each degree from 1° to 90°.

The modern name "tangent" did not make its debut until 1583, when Thomas Fincke (1561–1646), a Danish mathematician, used it in his *Geometria Rotundi*; up until then, most European writers still used terms taken from shadow reckoning: *umbra recta* ("straight shadow") for the horizontal shadow cast by a vertical gnomon, and *umbra versa* ("turned shadow") for a vertical shadow cast by a gnomon attached to a wall. The word "cotangens" [sic] was first used by Edmund Gunter in 1620. Various abbreviations were suggested for these functions, among them t and $t\,co$ by William Oughtred (1657) and T and t by John Wallis (1693). But the first to use such abbreviations consistently was Richard Norwood (1590–1665), an English mathematician and surveyor; in a work on trigonometry published in London in 1631 he wrote: "In these examples s stands for *sine*: t for *tangent*: sc for *sine complement* [i.e., cosine]: tc for *tangent complement*: sec for *secant*." We note that even today there is no universal conformity of notation, and European texts often use "tg" and "ctg" for tangent and cotangent.

The word "tangent" comes from the Latin *tangere*, to touch; its association with the tangent function may have come from the following observation: in a circle with center at O and radius r (fig. 16), let AB be the chord subtended by the central angle 2α, and OQ the bisector of this angle. Draw a line parallel to AB and tangent to the circle at Q, and extend OA and OB until they meet this line at C and D, respectively. We have

$$AB = 2r \sin \alpha, \quad CD = 2r \tan \alpha,$$

showing that the tangent function is related to the tangent line in the same way as the sine function is to the chord. Indeed, this construction forms the basis of the modern definition of the six trigonometric functions on the unit circle.

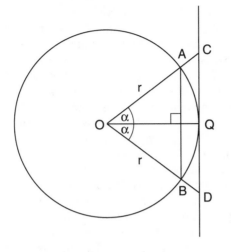

FIG. 16. $AB = 2r\sin\alpha$,
$CD = 2r\tan\alpha$.

✧ ✧ ✧

Through the Arabic translations of the classic Greek and Hindu texts, knowledge of algebra and trigonometry gradually spread to Europe. In the eighth century, Europe was introduced to the Hindu numerals—our modern decimal numeration system— through the writings of Mohammed ibn Musa al-Khowarizmi (ca. 780–ca. 840). The title of his great work, *'ilm al-jabr wa'l muqabalah* ("the science of reduction and cancelation") was transliterated into our modern word "algebra," and his own name evolved into the word "algorithm." The Hindu-Arabic system was not immediately accepted by the public, who preferred to cling to the old Roman numerals. Scholars, however, saw the advantages of the new system and advocated it enthusiastically, and contests between "abacists," who computed with the good old abacus, and "algorists," who did the same symbolically with paper and pen, became a common feature of medieval Europe.

It was mainly through Leonardo Fibonacci's exposition of the Hindu-Arabic numerals in his *Liber Abaci* (1202) that the decimal system finally took general hold in Europe. The first trigonometric tables using the new system were computed around 1460 by Georg von Peurbach (1423–1461). But it was his pupil Johann Müller (1436–1476), known as Regiomontanus (because he was born in Königsberg, which in German means "the royal mountain") who wrote the first comprehensive treatise on trigonometry up to date. In his *De triangulis omnimodis libri quinque* ("of triangles of every kind in five books," ca. 1464)[6] he developed the subject starting from basic geometric concepts and leading to the definition of the sine function; he then showed how to solve any triangle—plane or spherical—using either the sine of an angle or the sine of its complement (the cosine). The Law of Sines is stated here in verbal form, and so is the rule for finding the area of a triangle, $A = (ab \sin \gamma)/2$. Curiously the tangent function is absent, possibly because the main thrust of the work was spherical trigonometry where the sine function is dominant.

De triangulis was the most influential work on trigonometry of its time; a copy of it reached Copernicus, who studied it thoroughly (see p. 45). However, another century would pass before the word "trigonometry" appeared in a title of a book. This honor goes to Bartholomäus Pitiscus (1561–1613), a German clergyman whose main interest was mathematics. His book, *Trigonometriae sive de dimensione triangulorum libri quinque* (On trigonometry, or, concerning the properties of triangles, in five books), appeared in Frankfort in 1595. This brings us to the beginning of the seventeenth century, when trigonometry began to take the analytic character that it would retain ever since.

NOTES AND SOURCES

1. *The Aryabhatiya of Aryabhata: An Ancient Indian Work on Mathematics and Astronomy*, translated with notes by Walter Eugene Clark (Chicago: University of Chicago Press, 1930). In this work (p. 28) the value of π is given as 3.1416; this is stated in verbal form as a series of mathematical instructions, a common feature of Hindu mathematics. See also David Eugene Smith, *History of Mathematics*, (1925; rpt. New York: Dover, 1958), vol. 1, pp. 153–156, and George Gheverghese Joseph, *The Crest of the Peacock: Non-European Roots of Mathematics* (Harmondsworth, U.K.: Penguin Books, 1992), pp. 265–266.

2. In 1975 India named its first satellite after him.

3. For a detailed history of trigonometric notation, see Florian Cajori, *A History of Mathematical Notations* (1929; rpt. Chicago: Open Court, 1952), vol. 2, pp. 142–179; see also Smith, *History of Mathematics*, vol. 2, pp. 618–619 and 621-623. A list of trigonometric symbols, with their authors and dates, can be found in Vera Sanford, *A Short History of Mathematics* (1930; rpt. Cambridge, Mass.: Houghton Mifflin, 1958), p. 298.

4. Gauss-Schumacher correspondence, as quoted in Robert Edouard Moritz, *On Mathematics and Mathematicians (Memorabilia Mathematica)* (1914; rpt. New York: Dover, 1942), p. 318.

5. Smith, *History of Mathematics*, vol. 2, p. 620. Cajori, however, credits al-Battani as the first to construct a table of cotangents (*A History of Mathematics*, 1893; 2d ed. New York: Macmillan, 1919, p. 105).

6. English translation with an introduction and notes by Barnabas Hughes (Madison: University of Wisconsin Press, 1967).

Johann Müller, alias Regiomontanus

It is no coincidence that trigonometry up until the sixteenth century was developed mainly by astronomers. Aristarchus and Hipparchus, who founded trigonometry as a distinct branch of mathematics, were astronomers, as was Ptolemy, the author of the *Almagest*. During the Middle Ages, Arab and Hindu astronomers, notably Abul-Wefa, al-Battani, Aryabhata, and Ulugh Beg of Samarkand (1393–1449), absorbed the Greek mathematical heritage and greatly expanded it, especially in spherical trigonometry. And when this combined heritage was passed on to Europe, it was again an astronomer who was at the forefront: Johann Müller (see fig. 17).

Müller was born in Unfinden, near the town of Königsberg in Lower Franconia, in 1436 (this is not the more famous Königsberg—now Kaliningrad—in East Prussia). Different versions of his name survive: Johannes Germanus (because he was a German), Johannes Francus (because Franconia was also known as Eastern France), and Johann von Kunsperk, named after the town of Königsberg. Following the practice of scholars in his time he adopted a Latin name, Regio Monte, a literal translation of the German word "Königsberg" ("the royal mountain"), and in due time he was known as Regiomontanus. Even this name, however, exists in several versions: the French scientist Pierre Gassendi (Peter Gassendus, 1592–1655), who wrote the first definitive biography of Regiomontanus, refers to him as Joannes de Monte Regio, which has a distinctive French sound.[1]

Regiomontanus spent his early years studying at home. When he was twelve his parents sent him to Leipzig for his formal education, and after graduating he moved to Vienna, getting his bachelor's degree from its university at the age of fifteen. There he met the mathematician and astronomer Georg von Peurbach (1423–1461), with whom he formed a close friendship. Peurbach studied under Cardinal Nicholas of Cusa (1401–1464) but rejected the latter's speculation that the earth might be moving around the sun. He was an admirer of Ptolemy and was planning to publish a corrected version of the *Almagest*, based on existing Latin translations. He also undertook the preparation of a new and more accurate table of sines, using the re-

FIG. 17. A sketch of Regiomontanus.

cently adopted Hindu-Arabic numerals. The young Regiomontanus quickly came under Peurbach's influence, and the bonds between them reached the intensity of a father-to-son relationship. But then Peurbach suddenly died, not yet 38 years old. His untimely death left his plans unfulfilled and his pupil in a shock.

On his deathbed Peurbach entrusted his young student to complete the translation of the *Almagest*. "This became a sacred trust for the fatherless student," wrote Gassendi in his biography of Regiomontanus.[2] Regiomontanus completely dedicated himself to the task, learning Greek so that he could read Ptolemy in the original. In the course of his work he became interested in old Greek and Latin manuscripts and acquired them wherever he went; among his prizes was an incomplete manuscript of Diophantus which he discovered in 1464. He courted the friendship of many scholars, among them a Cretan, George of Trebizond (Georgius Trapezuntios, 1396-1486), an authority on Ptolemy who translated the *Almagest* and Theon's commentaries on it into Latin. The friendship, however, turned sour when Regiomontanus criticized Trebizond for serious errors in his interpretation of the commentaries, calling him "the most impudently perverse blabber-mouth."[3] These words, by one account, would have dire consequences.

His many travels brought him to Greece and Italy, where he visited Padua, Venice, and Rome. It was in Venice, in 1464, that he finished the work for which he is best remembered, *On Triangles of Every Kind* (see below). In addition to all these activities, Regiomontanus was also a practicing astrologer, seeing no contradiction between this activity and his scientific work (the great astronomer Johannes Kepler would do the same two centuries later). Around 1467 he was invited by King Mathias Huniades Corvinus of Hungary to serve as librarian of the newly founded royal library in Budapest; the king, who had just returned victorious from his war with the Turks and brought back with him many rare books as booty, found in Regiomontanus the ideal man to be in charge of these treasures. Shortly after Regiomontanus's arrival the king became ill, and his advisers predicted his imminent death. Regiomontanus, however, used his astrological skills to "diagnose" the illness as a mere heart weakness caused by a recent eclipse! Lo and behold, the king recovered and bestowed on Regiomontanus many rewards.

Regiomontanus returned to his homeland in 1471 and settled in Nürnberg, close to his birthplace. This city, known for its long tradition of learning and trade, had just opened a printing press, and Regiomontanus saw the opportunities it offered to spread the written word of science (it was just a few years earlier, in 1454, that Johann Gutenberg had invented printing from movable type). He founded his own press and was about to embark on an ambitious printing program of scientific manuscripts, but these plans were cut short by his early death. He also founded an astronomical observatory and equipped it with the finest instruments the famed Nürnberg artisans could produce; these included armillary spheres and devices for measuring angular distances between celestial objects.

Regiomontanus was the first publisher of mathematical and astronomical books for commercial use. In 1474 he printed his *Ephemerides*, tables listing the position of the sun, moon, and planets for each day from 1475 to 1506. This work brought him great acclaim; Christopher Columbus had a copy of it on his fourth voyage to the New World and used it to predict the famous lunar eclipse of February 29, 1504. The hostile natives had for some time refused Columbus's men food and water, and he warned them that God would punish them and take away the moon's light. His admonition was at first ridiculed, but when at the appointed hour the eclipse began, the terrified natives immediately repented and fell into submission.

In 1475 Pope Sixtus IV called upon Regiomontanus to come to Rome and help in revising the old Julian calendar, which was badly out of tune with the seasons. Reluctantly, he left his many duties and traveled to the Eternal City. And there, on July

6, 1476, he suddenly died, one month past his fortieth birthday. The cause of his death is not known: some blamed it on a plague, others on a passing comet. There were, however, persisting rumors that Regiomontanus was poisoned by Trebizond's sons, who had never forgotten his stinging criticism of their father's translation of Theon's commentaries on Ptolemy.[4] When news of his death became known, Nürnberg went into public mourning.[5]

✧ ✧ ✧

Regiomontanus's most influential work was his *De triangulis omnimodis* (On triangles of every kind), a work in five parts ("books") modeled after Euclid's *Elements* (fig. 18). In this work he organized into a systematic body of knowledge the trigonometric heritage of Ptolemy and the Hindu and Arab scholars. Book I begins with the definitions of basic concepts: quantity, ratio, equality, circle, arc, and chord. The sine function is introduced according to the Hindu definition: "When the arc and its chord are bisected, we call that half-chord the *right sine* of the half-arc." Next comes a list of axioms, followed by fifty-six theorems dealing with geometric solutions of plane triangles. Much of this material deals with geometry rather than trigonometry, but Theorem 20 introduces the use of the sine function in solving a right triangle.

Trigonometry proper begins in Book II with the enunciation of the Law of Sines; as with all other rules, this is stated literally rather than in symbols, but the formulation is as clear as in any present-day textbook. The sine law is then used to solve the cases *SAA* and *SSA* of an oblique triangle. Here also appears for the first time, though implicitly, the formula for the area of a triangle in terms of two sides and the included angle: "If the area of a triangle is given together with the rectangular product of the two sides, then either the angle opposite the base becomes known, or [that angle] together with [its] known [exterior] equals two right angles."[6] In modern formulation this says that from the formula $A = (bc \sin \alpha)/2$ one can find either α or $(180° - \alpha)$ if the area A and the product of two sides b and c are given. Strangely, Regiomontanus never uses the tangent function, although he must have been familiar with it from Peurbach's table of tangents of 1467, and of course from the Arabs' use of it in connection with shadow reckoning.[7]

The remaining three books deal with spherical geometry and trigonometry, both necessary tools in astronomy. As he states in his introduction, Regiomontanus's main goal in *On Triangles*

DOCTISSIMI VIRI ET MATHE-
maticarum diſciplinarum eximij profeſſoris

IOANNIS DE RE-

GIO MONTE DE TRIANGVLIS OMNI-
MODIS LIBRI QVINQVE:
Quibus explicantur res neceſſariæ cognitu, uolentibus ad
ſcientiarum Aſtronomicarum perfectionem deueni-
re:quæ cum nuſquã alibi hoc tempore expoſitæ
habeantur, fruſtra ſine harum inſtructione
ad illam quiſquam aſpirarit.

Acceſſerunt huc in calce pleraqʒ D. Nicolai Cuſani de Qua
dratura circuli, Deqʒ recti ac curui commenſuratione:
itemqʒ Io. de monte Regio eadem de re ἰ̉λϳϰ̓ϰ-
ϰά, hactenus à nemine publicata.

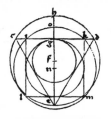

Omnia recens in lucem edita, fide & diligentia
ſingulari, Norimbergæ in ædibus Io. Petrei,
ANNO CHRISTI
M. D. XXXIII.

FIG. 18. Title page
of *On Triangles*
of *Every Kind*
(Nürnberg, 1533).

was to provide a mathematical introduction to astronomy. In words that might have been taken from a modern textbook, he admonishes his readers to study the book carefully, as its subject matter is a necessary prerequisite to an understanding of the heavens:

> You, who wish to study great and wonderous things, who wonder about the movement of the stars, must read these theorems about triangles. . . . For no one can bypass the science of triangles and reach a satisfying knowledge of the stars. . . . A new student should neither be frightened nor despair. . . . And where a theorem may present some problem, he may always look down to the numerical examples for help.[8]

Regiomontanus completed writing *On Triangles* in 1464, but it was not published until 1533, more than half a century after his death. Through George Joachim Rhæticus (1514–1576), the leading mathematical astronomer in Germany during the first half of the sixteenth century, it reached Nicolas Copernicus (1473–1543). Rhæticus visited Copernicus in 1539 and became his first disciple; though Rhæticus was younger than Copernicus

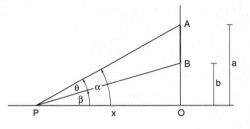

FIG. 19.
Regiomontanus's maximum problem: at what distance x will segment AB subtend the largest angle θ?

by forty-one years, the two studied together, with the former instructing the latter in mathematics. (It was at Rhæticus's persistent prodding that Copernicus finally agreed to publish his great work, *De revolutionibus*, in which he expounded his heliocentric system.) He presented Copernicus with an inscribed copy of *On Triangles*, which the great master thoroughly studied; this copy survived and shows numerous marginal notes in Copernicus's handwriting.[9] Later, Tycho Brahe (1546–1601), the great Danish observational astronomer, used the work as the basis for calculating the position of the famous nova (new star) in Cassiopeia, whose appearance in 1572 he was fortunate to witness. Regiomontanus's work thus laid much of the mathematical groundwork that helped astronomers shape our new view of the universe.

✧ ✧ ✧

In 1471 Regiomontanus posed the following problem in a letter to Christian Roder, a professor at the university of Erfurt: "At what point on the ground does a perpendicularly suspended rod appear largest [i.e., subtends the greatest visual angle]?" It has been claimed that this was the first extreme problem in the history of mathematics since antiquity.[10]

In figure 19 let the rod be represented by the vertical line segment AB. Let $OA = a$, $OB = b$, and $OP = x$, where P is the point on the ground at which the angle $\theta = \angle BPA$ is maximum. Let $\alpha = \angle OPA$, $\beta = \angle OPB$. We have

$$\cot \theta = \cot (\alpha - \beta) = \frac{\cot \alpha \cot \beta + 1}{\cot \beta - \cot \alpha}$$

$$= \frac{(x/a)(x/b) + 1}{x/b - x/a}$$

$$= \frac{x}{a - b} + \frac{ab}{(a - b)x}.$$

We might now be inclined to differentiate this expression in order to find the value of x that minimizes it (since $\cot \theta$ is decreas-

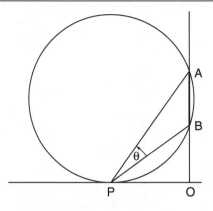

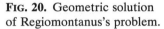

FIG. 20. Geometric solution of Regiomontanus's problem.

ing for $0° < \theta < 90°$, a maximum value of θ means a minimum value of cot θ). But Regiomontanus lived two hundred years before the invention of calculus, so let us restrict ourselves to elementary methods. We will use a theorem from algebra that says: the arithmetic mean of two positive numbers u and v is never smaller than their geometric mean, and the two means are equal if and only if the two numbers are equal. In symbols, $(u + v)/2 \geq \sqrt{uv}$, with equality holding if and only if $u = v$.[11] Putting $u = x/(a - b)$, $v = ab/(a - b)x$, we have

$$\cot \theta = u + v \geq 2\sqrt{uv}$$

$$= 2\sqrt{\frac{x}{a - b} \cdot \frac{ab}{(a - b)x}} = \frac{2\sqrt{ab}}{a - b},$$

with equality holding if and only if $x/(a - b) = ab/(a - b)x$, that is, when $x = \sqrt{ab}$. The required point is thus located at a distance equal to the geometric mean of the altitudes of the upper and lower endpoints of the rod, measured horizontally from the foot of the rod.

This result provides an interesting geometric interpretation: Using a straightedge and compass, construct the circle passing through A and B and tangent to the ground (fig. 20). By a well-known theorem, we have $OA \cdot OB = (OP)^2$, that is, $ab = x^2$ and therefore $x = \sqrt{ab}$. Conversely, we can easily convince ourselves that the circle passing through A, B, and the required point P must be tangent to the ground; for had it intercepted the ground at two points R and S (fig. 21), then the subtended angle at any point P between R and S would be greater than the angle at R or at S (P now being an interior point of the circle), whereas the angle at P was supposed to be the greatest. Thus Regiomontanus's problem can be solved by a simple geometric construction.[12]

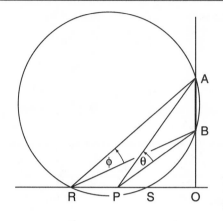

FIG. 21. Proof of the geometric solution.

We can only speculate what prompted Regiomontanus to propose this problem. It may have originated as a problem in architecture or perspective: to find the most favorable position from which to view a window of a tall building. Perspective—the technique of drawing objects according to their true appearance to the eye—was then a novel subject, introduced by the Italian Renaissance artists Filippo Brunelleschi (1377–1446) and Leone Battista Alberti (1404–1472). (The famed artist Albrecht Dürer, a major writer on perspective, was born in Nürnberg the year Regiomontanus settled there.) It soon became a central doctrine in art, and its juxtaposition of two seemingly alien disciplines— art and geometry—was in tune with the Renaissance ideal of universalism. It may well be that Regiomontanus posed his problem in response to an inquiry from an artist or architect.

NOTES AND SOURCES

1. See *Regiomontanus on Triangles*, translated by Barnabas Hughes with an introduction and notes (Madison: University of Wisconsin Press, 1967), pp. 11–17, from which this biographical sketch is adapted. See also David Eugene Smith, *History of Mathematics* (1923; rpt. New York: Dover, 1958), vol. 1, pp. 259–260. The only modern biography of Regiomontanus is in German: Ernst Zinner, *Leben und Wirken des Johannes Müller von Königsberg genannt Regiomontanus* (Munich: C. H. Beck, 1939).

2. As quoted in Hughes, *Regiomontanus on Triangles*, p. 13.

3. Ibid., p. 14.

4. The story about his death by poisoning is told in Gassendi's biography of Regiomontanus, dated 1654, and evokes echoes of Antonio Salieri allegedly poisoning his arch rival, Mozart.

5. The French humanist Peter Ramus (Pierre de la Ramée, 1515–1572) attributed to Regiomontanus the invention of a mechanical fly that could leave a person's hand, buzz about the room, and then return to the person's hand, and an eagle that could leave the city, fly to meet an approaching dignitary, and then accompany him back to the city. (Hughes, *Regiomontanus on Triangles*, p. 17). These stories are no doubt an exaggeration, but they reflect the high esteem with which Regiomontanus was held by his fellow townsmen. In Ramus's words, "Tarento had its Archytas, Syracuse its Archimedes, Byzantium Proclus, Alexandria Ctesibius, and Nürnberg Regiomontanus. . . . The mathematicians of Tarento and Syracuse, of Byzantium and Alexandria are gone. But among the Masters of Nürnberg, the joy of the scholars is the mathematician Regiomontanus."

6. Ibid., p. 133.

7. Ibid., pp. 4–7.

8. Ibid., pp. 27–29.

9. Ibid., p. 9.

10. Heinrich Dörrie, *100 Great Problems of Elementary Mathematics: Their History and Solution*, translated by David Antin (1958; rpt. New York: Dover, 1965), pp. 369–370. I have slightly changed the wording so as to make the reading easier.

11. This theorem follows from the fact that the square of a real number can never be negative; hence $0 \leq (\sqrt{u} - \sqrt{v})^2 = u - 2\sqrt{uv} + v$. Moving the term $-2\sqrt{uv}$ to the left side and dividing by 2, we get the required result. Equality occurs if and only if $\sqrt{u} - \sqrt{v} = 0$, i.e., $u = v$.

12. I was unable to find out if Regiomontanus actually provided a solution. According to Florian Cajori (*A History of Mathematical Notation*, 1928; rpt. La Salle, Ill.: Open Court, 1951, vol. 1, p. 95), Regiomontanus's correspondence with fellow scientists during the period 1463–1471 is preserved at the Stadtbibliothek of Nürnberg.

4

Trigonometry Becomes Analytic

Thus the analysis of angular sections involves geometric
and arithmetic secrets which hitherto have been
penetrated by no one.
—François Viète

With the great French mathematician François Viète (also known by his Latin name Franciscus Vieta, 1540–1603), trigonometry began to assume its modern, analytic character. Two developments made this process possible: the rise of symbolic algebra—to which Viète was a major contributor—and the invention of analytic geometry by Fermat and Descartes in the first half of the seventeenth century. The gradual replacement of the cumbersome verbal algebra of medieval mathematics with concise, symbolic statements—a literal algebra—greatly facilitated the writing and reading of mathematical texts. Even more important, it enabled mathematicians to apply algebraic methods to problems that until then had been treated in a purely geometric way.

With Viète, trigonometry underwent a second important change: it admitted infinite processes into its ranks. In 1593 he discovered the famous infinite product

$$\frac{2}{\pi} = \frac{\sqrt{2}}{2} \cdot \frac{\sqrt{2+\sqrt{2}}}{2} \cdot \frac{\sqrt{2+\sqrt{2+\sqrt{2}}}}{2} \cdots$$

(Viète used the abbreviation *etc.* instead of the three dots). It was the first time an infinite process was explicitly written as a mathematical formula, and it heralded the beginning of modern analysis.[1] (We will prove Viète's product in chapter 11.)

In England three individuals made substantial contributions to trigonometry in the first half of the seventeenth century. John Napier's (1550–1617) invention of logarithms in 1614 enormously aided numerical calculations, particularly in trigonometry.[2] William Oughtred (1574–1660) was the first to attempt a systematic use of trigonometric symbols: in his work *Trigonometrie, or, The manner of Calculating the Sides and Angles of Triangles, by the Mathematical Canon, demonstrated* (London, 1657), he used the abbreviations *s*, *t*, *se*, *s co*, *t co*, and *se co* for sine, tangent, secant, cosine ("sine complement"), cotangent, and cosecant, respectively.[3] (In contrast to its long title, the work itself contains, besides the tables, only thirty-six pages of text.) And John Wallis's (1616–1703) work on infinite series was an immediate forerunner to Newton's discoveries in the same field. Wallis, more than anyone else up to his time, realized that synthetic methods in mathematics should give way to analytic ones: he was the first to treat conic sections as quadratic equations rather than geometric objects, as the Greeks had done. (Wallis was also the first major mathematician to write on the history of mathematics, and he introduced the symbol ∞ for infinity.) His most famous formula is the infinite prduct

$$\frac{\pi}{2} = \frac{2}{1} \cdot \frac{2}{3} \cdot \frac{4}{3} \cdot \frac{4}{5} \cdot \frac{6}{5} \cdot \frac{6}{7} \cdots,$$

which together with Viète's product ranks among the most beautiful formulas in mathematics. Wallis arrived at this result by daring intuition and a complex interpolation process that would stretch the patience of a modern reader to the limit;[4] in chapter 12 we will derive his product in a shorter and more elegant way.

❖ ❖ ❖

There was yet another reason for the rise of analytic trigonometry in the first half of the seventeenth century: the ever increasing role of mathematics in describing the physical world around us. Whereas the inventors of classical trigonometry were mainly interested in applying it to the heavens (hence the initial predominance of spherical over plane trigonometry), the new era had its feet firmly planted in the mechanical world of daily life. Galileo's discovery that every motion can be resolved into two components along perpendicular lines—and that these components can be treated independently of each other—at once made trigonometry indispensable in the study of motion. The science of artillery—and in the seventeenth century it *was* regarded as a science—was chiefly concerned with finding the range of a projectile fired from a cannon. This range, in the absence of air resistance, is given by the formula $R = (v_0^2 \sin 2\alpha)/g$, where v_0 is the velocity of the projectile as it leaves the cannon (the muzzle velocity), α the angle of firing relative to the ground, and g

the acceleration due to gravity (about 9.81 m/sec^2). This formula shows that for a given velocity, the range depends solely on α: it reaches its maximum when $\alpha = 45°$ and falls off symmetrically on either side of $45°$. These facts, of course, had been known empirically for a long time, but their theoretical basis was new in Galileo's time.

Another branch of mechanics vigorously studied in the seventeenth and eighteenth centuries dealt with oscillations. The great sea voyages of the era demanded ever more accurate navigational techniques, and these in turn depended on the availability of clocks of ever greater precision. This led scientists to study the oscillations of pendulums and springs of various kinds. Some of the greatest names of the day were involved in these studies, among them Christiaan Huygens (1629–1695) and Robert Hooke (1635–1703). Huygens discovered the cycloidal pendulum, whose period of oscillation is independent of the amplitude, while Hooke's work on the coiled spring laid the basis for the modern spring-driven watch. On another level, the increased skill and sophistication in building musical instruments—from woodwinds and brass to keyboard instruments and organs—motivated scientists to study the vibrations of sound-producing bodies such as strings, membranes, bells, and air pipes. All this underscored the role of trigonometry in describing *periodic phenomena* and resulted in a shift of emphasis from computational trigonometry (the compilation of tables) to the *relations* among trigonometric functions—the essence of analytic trigonometry.

✧ ✧ ✧

In his work *Harmonia mensurarum* (Harmony of mensuration), published posthumously in 1722, the English mathematician Roger Cotes (1682–1716) gave the equivalent of the formula (in modern notation)

$$\phi i = \log(\cos \phi + i \sin \phi),$$

where $i = \sqrt{-1}$ and "log" stands for natural logarithm (logarithm to the base $e = 2.718\ldots$). This, of course, is equivalent to Leonhard Euler's famous formula $e^{i\phi} = \cos \phi + i \sin \phi$, published in 1748 in his great work, *Introductio in analysin infinitorum*. Also in 1722 Abraham De Moivre (1667–1754) derived—though in implicit form—the formula,

$$(\cos \phi + i \sin \phi)^n = \cos n\phi + i \sin n\phi,$$

which is the basis for finding the nth root of a number, real or complex. It took, however, the authority of Euler and his *Introductio* to fully incorporate complex numbers into trigonometry: with him the subject became truly analytic. (We will return to the role of complex numbers in trigonometry in chapter 14.)

These developments moved trigonometry ever farther from its original connection with a triangle. The first to define the trigonometric functions as pure numbers, rather than ratios in a triangle, was Abraham Gotthelf Kästner (1719–1800) of Germany; in 1759 he wrote: "If x denotes the angle expressed in degrees, then the expressions sin x; cos x; tang x etc. are numbers, which correspond to every angle."[5] Today, of course, we go a step farther and define the independent variable itself as a real number rather than an angle.

✧   ✧   ✧

Almost from its inception the differential and integral calculus was applied to numerous problems in mechanics, first to discrete mechanics (the motion of a single particle or a system of particles), and later to continuous mechanics. Among the latter, the outstanding problem in the second half of the eighteenth century was the vibrating string. This problem has excited mathematicians since the earliest time because of its close association with music. Already Pythagoras, in the sixth century B.C., discovered some of the laws governing the vibrations of a string; this led him to construct a musical scale based on mathematical principles. A full investigation of the problem, however, required methods that were not even available to Newton and Leibniz, namely partial differential equations (equations in which the unknown function and its derivatives depend on two or more independent variables). For the vibrating string the relevant equation is $\partial^2 u/\partial x^2 = (1/c^2)(\partial^2 u/\partial t^2)$, where $u = u(x, t)$ is the displacement from equilibrium of a point at a distance x from one endpoint of the string at time t, and c is a constant that depends on the physical parameters of the string (its tension and linear density).

The attempts to solve this famous equation, known as the one-dimensional wave equation, involved the best mathematical minds of the time, among them the Bernoulli family, Euler, D'Alembert, and Lagrange. Euler and D'Alembert expressed their solutions in terms of arbitrary functions representing two waves, one moving along the string to the right, the other to the left, with a velocity equal to the constant c. Daniel Bernoulli, on the other hand, found a solution involving an infinite series of trigonometric functions. Since these two types of solution to the same problem looked so different, the question arose whether they could be reconciled, and if not, to find out which was the more general. This question was answered by the French mathematician Jean Baptiste Joseph Fourier (1768–1830). In his most important work, *Théorie analytique de la chaleur* (Analytic theory of heat, 1822), Fourier showed that almost any function, when regarded as a periodic function over a given interval, can be represented by a trigonometric series of the form

$$f(x) = a_o + a_1 \cos x + a_2 \cos 2x + a_3 \cos 3x + \cdots$$
$$+ b_1 \sin x + b_2 \sin 2x + b_3 \sin 3x + \cdots,$$

where the coefficients a_i and b_i can be found from $f(x)$ by computing certain integrals. This *Fourier series* is in some respects more general than the familiar Taylor expansion of a function in a power series; for example, while the Taylor series can be applied only to functions $f(x)$ that are continuous and have continuous derivatives, the Fourier series may exist even if $f(x)$ is discontinuous. We will return to these series in chapter 15.

Fourier's theorem marks one of the great achievements of nineteenth-century analysis. It shows that the sine and cosine functions are essential to the study of *all* periodic phenomena, simple or complex; they are, in fact, the building blocks of all such phenomena, in much the same way that the prime numbers are the building blocks of all integers. Fourier's theorem was later generalized to nonperiodic functions (in which case the infinite series becomes an integral), as well as to series involving nontrigonometric functions. These developments proved to be of crucial importance to numerous branches of science, from optics and acoustics to information theory and quantum mechanics.

NOTES AND SOURCES

1. Recent findings indicate that the Hindus may have known several infinite series involving π before Viète; see George Gheverghese Joseph, *The Crest of the Peacock: Non-European Roots of Mathematics* (Harmondsworth, U.K.: Penguin Books, 1991), pp. 286–294.

2. See my book, *e: The Story of a Number* (Princeton, N.J.: Princeton University Press, 1994), chaps. 1 and 2.

3. See, however, pp. 37–38 as to the priority of using these symbols. See also Florian Cajori, *William Oughtred: A Great Seventeenth-Century Teacher of Mathematics* (Chicago: Open Court, 1916), pp. 35–39. Cajori notes that the tables in Oughtred's book use a centesimal division of the

degree (i.e., into one hundred parts), a practice that has been revived in our time with the advent of the hand-held calculator.

4. See *A Source Book in Mathematics, 1200–1800* ed. D. J. Struik (Cambridge, Mass.: Harvard University Press, 1969), pp. 244–253.

5. David Eugene Smith, *History of Mathematics* (1925; rpt. New York: Dover, 1958), vol. 2, p. 613. Kästner was the first mathematician to write a work entirely devoted to the history of mathematics (in 4 vols.; Göttingen, 1796–1800).

François Viète

It is unfortunate that the names of so many of those who helped shape mathematics into its present form have largely vanished from today's curriculum. Among them we may mention Regiomontanus, Napier, and Viète, all of whom made substantial contributions to algebra and trigonometry.

François Viète was born in Fontenay le Comte, a small town in western France, in 1540 (the exact day is unknown). He first practiced law and later became involved in politics, serving as member of the parliament of Brittany. As was the practice of many learned men at the time, he latinized his name to Franciscus Vieta; but unlike others—among them Regiomontanus—the Latin version was not adopted universally. We shall use the French Viète.

Throughout his life Viète practiced mathematics only in his leisure time, regarding it as an intellectual recreation rather than a profession. He was not alone in this attitude: Pierre Fermat, Blaise Pascal, and René Descartes all made great contributions to mathematics at their leisure while officially occupying a variety of political, diplomatic, and, in Descartes' case, military positions. Viète began his scientific career as a tutor to Catherine of Phartenay, the daughter of a prominent military figure, for whom he wrote several textbooks. As his reputation grew, he was called on to serve the monarch Henry IV in his war against Spain. Viète proved himself an expert in breaking the enemy code: a secret message to the Spanish monarch Philip II from his liaison officer was intercepted by the French and given to Viète, who succeded in deciphering it. The Spaniards, amazed that their code had been broken, accused the French of using sorcery, "contrary to the practice of the Christian faith."[1]

Viète's most important work is his *In artem analyticam isagoge* (Introduction to the analytical art; Tours, 1591), considered the earliest work on symbolic algebra. In this work he introduced a system of notation that came close to the one we use today: he denoted known quantities by consonants and unknowns by vowels. (The present custom of using *a*, *b*, *c*, etc. for constants and *x*, *y*, *z* for unknowns was introduced by Descartes in 1637.) He defined an equation as "a comparison of an unknown magnitude with a determinate one" and gave the basic rules for solving equations—moving a term from one side of the equation to the other, dividing an equation by a common factor, and so on. He called his method *ars analytice* and the new algebra *logistica speciosa* (literally, the art of calculating with species, i.e., general quantities), to distinguish it from the old *logistica numerosa*. This transition from verbal to symbolic algebra is considered one of the most important developments in the history of mathematics.

Viète applied the rules of algebra to *any* quantity, arithmetic or geometric, thereby putting to rest the age-old distinction between pure numbers and geometric entities. In other respects, however, he was rather conservative. For example, he always insisted on making an equation dimensionally homogeneous: instead of the modern equation $mx = b$ he would write "*M in A aequatur B quadratus*", meaning that a given quantity M (represented by a consonant) multiplied by an unknown quantity A (a vowel) is equal to the square of a given number B. This shows that he was still clinging to the old Greek view that regarded operations among numbers as geometric in nature; since a product of two numbers represents the area of a rectangle having these numbers as sides, one must equate such a product to the area of a square whose side is given. (Today, of course, we regard algebraic quantities as pure, dimensionless numbers.) It is also interesting to note that Viète used the modern symbols $+$ and $-$ for addition and subtraction, but for equality he used the verbal description "aequatur." For A^2 he wrote *A quadratus* and for A^3, *A cubus* (although later he abbreviated these to Aq and Ac). Clearly Viète could not entirely free himself from the shackles of the old verbal algebra. His work reflects the time in which he lived—a period of transition from the old world to the new.

❖   ❖   ❖

Of particular interest to us are Viète's contributions to trigonometry. His first work on this subject appeared in 1571 under the title *Canon mathematicus seu ad triangula cum appendicibus*. Here he gives the first systematic treatment in the Western world of the methods for solving plane and spherical triangles, using all six trigonometric functions. He develops the three sum-to-product formulas (e.g., $\sin\alpha + \sin\beta = 2\sin(\alpha+\beta)/2 \cdot \cos(\alpha-\beta)/2$, with similar formulas for $\sin\alpha + \cos\beta$ and $\cos\alpha + \cos\beta$), from which Napier may have gotten the idea of logarithms, since they allow one (when used in reverse) to reduce a product of two numbers to the sum of two other numbers. And he was the first to state the Law of Tangents in its modern form: $(a+b)/(a-b) = [\tan(\alpha+\beta)/2]/[\tan(\alpha-\beta)/2)]$, where a and b are two sides of a triangle and α and β the opposite angles (see page 152).

Viète was the first mathematician to systematically apply algebraic methods to trigonometry. For example, by letting $x = 2\cos\alpha$ and $y_n = \cos n\alpha$, he obtained the recurrence formula

$$y_n = xy_{n-1} - y_{n-2},$$

which, when translated back into trigonometry, becomes

$$\cos n\alpha = 2\cos\alpha \cdot \cos(n-1)\alpha - \cos(n-2)\alpha.$$

One can now express $\cos(n-1)\alpha$ and $\cos(n-2)\alpha$ in terms of the cosines of still lower multiples of α; and by continuing the process, one ends up with a formula expressing $\cos n\alpha$ in terms of $\cos\alpha$ and $\sin\alpha$. Viète was able to do this for all integers n up to 10. He was so proud of his achievement that he exclaimed, "Thus the analysis of angular sections involves geometric and arithmetic secrets which hitherto have been penetrated by no one."[2] To appreciate his feat, we mention that the general formulas expressing $\cos n\alpha$ and $\sin n\alpha$ in terms of $\cos\alpha$ and $\sin\alpha$ were found by Jakob Bernoulli in 1702, more than a hundred years after Viète's work.[3]

Viète's adeptness in applying algebraic transformations to trigonometry served him well in a famous encounter between Henry IV and Netherland's ambassador to France. Adriaen van Roomen (1561–1615), professor of mathematics and medicine in Louvain (Belgium), in 1593 published a work entitled *Ideae mathematicae*, which contained a survey of the most prominent living mathematicians of the time.[4] Not a single French mathematician was mentioned, prompting the Dutch ambassador to speak disdainfully about France's scientific achievements. To prove his point, he presented Henry IV with a problem that appeared in the *Ideae*—with a prize offered to whoever will solve it—and boasted that surely no French mathematician will come up with a solution. The problem was to solve the 45th-degree equation

$$x^{45} - 45x^{43} + 945x^{41} - 12{,}300x^{39} + - \cdots$$
$$+ 95{,}634x^5 - 3{,}795x^3 + 45x = c,$$

where c is a constant.

Henry summoned Viète, who immediately found one solution, and on the following day came up with twenty-two more.

What happened is described by Florian Cajori in *A History of Mathematics*:

Viète, who, having already pursued similar investigations, saw at once that this awe-inspiring problem was simply the equation by which $c = 2 \sin \phi$ was expressed in terms of $x = 2 \sin (\phi/45)$; that, since $45 = 3 \cdot 3 \cdot 5$, it was necessary only to divide an angle once into five equal parts, and then twice into three—a division which could be effected by corresponding equations of the fifth and third degrees.[5]

To follow Viète's line of thought, let us first look at a simpler problem. Suppose we are asked to solve the equation

$$x^3 - 3x + 1 = 0.$$

We rewrite it as $1 = 3x - x^3$ and make the substitution $x = 2y$:

$$\frac{1}{2} = 3y - 4y^3.$$

If we have a keen eye, we might recognize the similarity between this equation and the identity

$$\sin 3\alpha = 3 \sin \alpha - 4 \sin^3 \alpha.$$

In fact, we can make the two equations coincide if we write $1/2 = \sin 3\alpha$ and $y = \sin \alpha$; in this new form, the problem amounts to finding $\sin \alpha$, given that $\sin 3\alpha = 1/2$. But if $\sin 3\alpha = 1/2$, then $3\alpha = 30° + 360°k$, where $k = 0, \pm 1, \pm 2, \ldots$, and so $\alpha = 10° + 120°k$. Hence $y = \sin (10° + 120°k)$, and finally $x = 2y = 2 \sin (10° + 120°k)$. However, because the sine function has a period of $360°$, it suffices to consider only the values $k = 0, 1, 2$. Our three solutions are then

$$x_0 = 2 \sin 10° = 0.347 \cdots,$$

$$x_1 = 2 \sin 130° = 1.532 \cdots,$$

and

$$x_2 = 2 \sin 250° = -1.879 \cdots.$$

With a calculator we can easily check that these are indeed the three solutions. Thus a trigonometric identity helped us to solve a purely algebraic equation.

Now it is one thing to solve a cubic equation using trigonometry, but quite another to solve an equation of degree 45. How then did Viète find his solutions? In a work entitled *Responsum* (1595) he outlined his method, which we summarize here in modern notation: let

$$c = 2 \sin 45\theta, \; y = 2 \sin 15\theta, \; z = 2 \sin 5\theta, \; x = 2 \sin \theta.$$

Our task is to find $x = 2\sin\theta$, given that $c = 2\sin 45\theta$. We will do this in three stages. We again start with the identity $\sin 3\alpha = 3\sin\alpha - 4\sin^3\alpha$. Multiplying by 2 and susbtituting first $\alpha = 15\theta$, we get

$$c = 3y - y^3. \tag{1}$$

Next, we substitute $\alpha = 5\theta$ and get

$$y = 3z - z^3. \tag{2}$$

We now use the identity

$$\sin^5\alpha = \frac{5}{8}\sin\alpha - \frac{5}{16}\sin 3\alpha + \frac{1}{16}\sin 5\alpha.^6$$

Multiplying by 32, replacing α by θ and expressing $2\sin 3\theta$ in terms of $x = 2\sin\theta$, we get

$$x^5 = 10x - 5(3x - x^3) + z,$$

which after simplifying becomes

$$z = 5x - 5x^3 + x^5. \tag{3}$$

If we now back-substitute equation (3) into (2) and then (2) into (1) and expand, we get exactly van Roomen's equation!

Viète thus split the original problem into three simpler ones. But why did he find only twenty-three solutions, when we know that the original equation must have forty-five solutions (all real, as follows from the geometric meaning of the problem: to divide an arbitrary angle into forty-five equal parts)? The reason is that in Viète's time it was still the practice to regard the *chord length* of an angle, rather than its sine, as the basic trigonometric function (see chapter 2); and since length is non-negative, he had to reject all negative solutions as meaningless. The complete set of solutions is given by

$$x_k = 2\sin(\theta + 360°k/45), k = 0, 1, 2, \ldots, 44;$$

of these (assuming that $45\theta \le 180°$, for otherwise $\sin 45\theta$ itself would already be negative), only the first twenty-three are positive, corresponding to angles in quadrants I and II.

✧   ✧   ✧

Among Viète's many other contributions we should mention his discovery of the relation between the roots of a quadratic equation $ax^2 + bx + c = 0$ and its coefficients ($x_1 + x_2 = -b/a$ and $x_1 x_2 = c/a$), although his rejection of negative roots prevented him from stating this relation as a general rule; the development of a numerical method for approximating the solutions of algebraic equations; and his discovery of the famous infinite product for π that bears his name (see page 50). Most of his works were originally printed for private circulation only; they were collected, edited, and published in 1646, more than forty years after Viète's death, by the Dutch mathematician Frans van Schooten (1615–1660).[7]

During Viète's last years he was embroiled in a bitter controversy with the German mathematician Christopher Clavius (1537–1612) over the reformation of the calendar that had been ordered by Pope Gregory XIII in 1582. Viète's caustic attacks on Clavius, who was the pope's adviser in this matter, made him many enemies and resulted in his adversaries' rejection of his new algebra. It is also worth noting that Viète consistently opposed the Copernican system, attempting instead to improve the old geocentric system of Ptolemy. We see here the inner conflict of a man who was at once an innovator of the first rank and a conservative deeply rooted in the past. Viète died in Paris on December 13, 1603, at the age of sixty-three. With him, algebra and trigonometry began to take the form we know today.[8]

NOTES AND SOURCES

1. W. W. Rouse Ball, *A Short Account of the History of Mathematics* (1908; rpt. New York: Dover, 1960), p. 230.
2. Florian Cajori, *A History of Mathematics* (1893; 2nd ed. New York: MacMillan, 1919), p. 138.
3. These formulas are

$$\cos n\alpha = \cos^n \alpha - \frac{n(n-1)}{2!} \cos^{n-2} \alpha \cdot \sin^2 \alpha + \cdots$$

and

$$\sin n\alpha = \frac{n}{1!} \cos^{n-1} \alpha \cdot \sin \alpha - \frac{n(n-1)(n-2)}{3!} \cos^{n-3} \alpha \cdot \sin^3 \alpha + \cdots .$$

4. In this work the value of π is given to seventeen decimal places, a remarkable achievement at the time.
5. Cajori, *History of Mathematics*, p. 138.
6. This identity can be obtained from the formula $\sin 5\alpha = 5 \sin \alpha - 20 \sin^3 \alpha + 16 \sin^5 \alpha$ by replacing $\sin^3 \alpha$ with $(3 \sin \alpha - \sin 3\alpha)/4$ and solving for $\sin^5 \alpha$.
7. The van Schooten family produced three generations of mathematicians, all of whom were born and lived in Leyden: Frans senior (1581–1646), Frans junior mentioned above, and his half-brother Petrus (1634–1679). Of the three, the most prominent was Frans junior, who edited the Latin edition of Descartes's *La Géométrie*; he also wrote on perspective and advocated the use of three-dimensional coordinates

in space. He was the teacher of the great Dutch scientist Christiaan Huygens.

8. There is no biography of Viète in English. Sketches of his life and work can be found in Ball, *Short Account*, pp. 229–234; Cajori, *History of Mathematics*, pp. 137–139; Joseph Ehrenfried Hofmann, *The History of Mathematics*, translated from the German by Frank Gaynor and Henrietta O. Midonick (New York: Philosophical Library, 1957), pp. 92–101; and in the *DSB*.

5

Measuring Heaven and Earth

The science of trigonometry was in a sense a precursor
of the telescope. It brought faraway objects within the
compass of measurement and first made it possible
for man to penetrate in a quantitative manner the far
reaches of space.
—Stanley L. Jaki, *The Relevance of Physics* (1966).

Since its earliest days, geometry has been applied to practical problems of measurement—whether to find the height of a pyramid, or the area of a field, or the size of the earth. Indeed, the very word "geometry" derives from the Greek *geo* (earth) and *metron* (to measure). But the ambition of the early Greek scientists went even farther: using simple geometry and later trigonometry, they attempted to estimate the size of the universe.

Aristarchus of Samos (ca. 310–230 B.C.) is considered the first great astronomer in history. Whereas most of his predecessors had derived their picture of the cosmos from aesthetic and mythological principles, Aristarchus based his conclusions entirely on the observational data available to him. For example, he pointed out that the motion of the planets could best be accounted for if we assume that the sun, and not the earth, is at the center of the universe—this almost two thousand years before Copernicus proposed his heliocentric system.[1] Most of Aristarchus's writings are lost, but one work, *On the Sizes and Distances of the Sun and Moon*, a treatise on mathematical astronomy, has survived. In it he developed a geometric method for determining the ratio of the distances of the sun and moon from the earth.

His method, known as "lunar dichotomy" (from the Greek *dichotomos*, to divide into two parts), was based on the fact that at the moment when exactly half of the moon's disk appears to be illuminated by the sun, as happens twice during each lunar cycle, the lines of sight from the earth to the moon and from the moon to the sun form a right angle (fig. 22). It follows that if we know angle *MES*, we can, in principle, find the ratios of the sides of triangle *EMS*, and in particular the ratio *ES/EM*. Aristarchus says that ∠*MES* is "less than a quadrant by a thirtieth of a quadrant," that is, ∠*MES* = 90° − 3° = 87°. Using modern trigonometry, it then follows that *ES/EM* = sec 87° = 19.1. Aristarchus, of course, did not have trigonometric tables at his disposal, so he had to rely on a theorem that, in modern notation, says: if α and β are two acute angles and $\alpha > \beta$, then $(\sin \alpha)/(\sin \beta) < \alpha/\beta < (\tan \alpha)/(\tan \beta)$.[2] From this he concluded that *ES/EM* is greater than 18:1 but less than 20:1.

Now this estimate of the ratio *ES/EM* falls far short of the actual value, about 390. The reason is that Aristarchus's method, while sound in principle, was woefully impractical to implement. For one, it is extremely difficult to determine the exact moment of dichotomy, even with a modern telescope; and second, it is just as difficult to measure the angle *MES*—one must look directly into the sun, and the sun may have already set at the instant of dichotomy. Furthermore, because ∠*MES* is very close to 90°, a small error in its determination leads to a *large* error in the ratio *ES/EM*. For example, if ∠*MES* were 88° instead of 87°, *ES/EM* would be 28.7, while for 86° it would be 14.3. Still, Aristarchus's method marks the first attempt to estimate the dimensions of our planetary system based on an actual measurement of observable quantities.

Aristarchus also estimated the ratio of the *sizes* of the sun and moon. During a total eclipse of the sun, the moon completely covers the solar disk—but just barely so, which is why the duration of totality is so short—no longer than about seven minutes, and usually much less.[3] This means that the *apparent* sizes of the sun and moon as seen from the earth are roughly equal (about half a degree of arc as measured on the celestial sphere). Therefore the ratio of the actual diameters of these bodies must be nearly the same as the ratio of their distances from the earth.

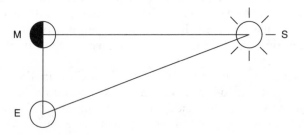

FIG. 22.
Aristarchus's
method.

Aristarchus thus concluded that the sun's diameter is between 18 and 19 times that of the moon. The actual ratio is about 400.

Now it is one thing to estimate the *ratio* of distances of two faraway objects, but quite another to estimate their *actual* distances and sizes. Here the phenomenon of *parallax* plays a crucial role. It is a common experience that an object appears to change its position—as viewed against a remote background—when the observer changes his own position, or when viewed simultaneously by two observers at different places. If one knows the distance between the two observers (the length of the baseline) then, by measuring the apparent angular shift in the position of the object (the angle of parallax), one can find the distance to the object, using simple trigonometry. The method of parallax is the basis of terrestrial surveying, but when applied to the vast distances between celestial bodies, its accuracy is limited: the farther the object, the smaller its angle of parallax and the greater the uncertainty in estimating its distance.

Because the moon is relatively close to us, its apparent shift in position as seen by two observers at different locations on the earth, though small by terrestrial standards, is considerable on an astronomical scale. Still, to be of any practical use, the two observers must be stationed as far from each other as possible—ideally at opposite points on the earth. However, during the rare occurrence of a total solar eclipse, even a slight change in the observer's position can mean the difference between total darkness and a partially eclipsed sun. This was dramatically shown during the eclipse of January 24, 1925, which passed right through New York City and was watched by millions of people in clear skies. To determine the exact edge of the moon's shadow, observers were stationed at every intersection between 72nd and 135th Streets in Manhattan and instructed to report whether they had seen the sun's corona—visible only during totality, when the solar disk is completely covered by the moon—or merely a narrow crescent sun, indicating that the eclipse was partial. "The results were definite: the edge of the umbra [the moon's shadow] passed between 95th and 97th Streets, yielding an accuracy of several hundred feet for a shadow cast at a distance of over 200,000 miles."[4]

The first to use lunar parallax to estimate the moon's distance was Hipparchus of Nicaea, whom we have already met in chapter 2. Hipparchus carefully studied ancient Babylonian eclipse records dating back to the eighth century B.C., and from these he gained a thorough understanding of the motion of the sun and moon. By a fortunate coincidence, a solar eclipse had occurred not far from his birthplace just a few years before he was born;

this eclipse, recently identified as that of March 14, 189 B.C., was total near the Hellespont (the straight of Dardanelles in modern Turkey), whereas in Alexandria only four-fifths of the sun's disk were hidden by the moon. Since the sun and moon subtend about half a degree of arc as measured on the celestial sphere, the moon's apparent shift in position amounted to one-fifth of this, or about 6 arc minutes. Combining this information with the longitude and latitude of the two places and the elevation of the sun and moon at the time of the eclipse, Hipparchus was able to calculate the moon's least and greatest distance as 71 and 83 earth radii, respectively. While these estimates are in excess of the modern values of 56 and 64, they came within the correct order of magnitude and should be considered a remarkable achievement for his time.[5]

✧   ✧   ✧

Hipparchus had estimated the moon's distance in terms of earth radii. To express this distance in more common units, one must know the size of the earth. The notion that the earth is spherical is attributed to Pythagoras; whether he derived this idea from observational evidence (for example, from the fact that during a partial lunar eclipse the earth always casts a circular shadow on the moon) or, as is more likely, from aesthetic and philosophical principles (the sphere being the most perfect of all shapes) is not known. But once the notion of a spherical earth took hold, attempts were made to determine its size. The credit for achieving this feat goes to a brilliant mathematician and geographer of the second century B.C., Eratosthenes of Cyrene (ca. 275–194 B.C.).

Eratosthenes was a friend of Archimedes, the greatest scientist of the ancient era, who addressed to him several of his works. As with most scholars in earlier times, Eratosthenes was active in several disciplines. He prepared a celestial map that included 675 stars, and he determined the angle of inclination of the equator to the ecliptic (the plane of earth's orbit around the sun)—about 23.5°. He suggested to add to the calendar one extra day every four years to keep it in step with the seasons, an idea on which the Julian calendar was later based. In mathematics he devised the famous "sieve" for finding the prime numbers, and he gave a mechanical solution to the duplication problem: to find the side of a cube whose volume equals twice that of a given cube. Eratosthenes also wrote poetry and literary critique, and was the first to undertake a scientific chronology of major historical events going back to the Trojan War. His friends nicknamed him "Beta," perhaps because they ranked him second to Archimedes; but this slight did not prevent Ptolemy III, the ruler of Egypt, to call upon him to head the great library of Alexandria, the largest repository of scholarly works in the ancient world. In old age he became blind, and sensing that his productive years were over, died a "philosopher's death" by voluntary starvation.

In the year 240 B.C. Eratosthenes achieved the feat for which he is chiefly remembered: computing the size of the earth. It was known that at noon on the day of the summer solstice (the longest day of the year), the sun's rays directly illuminated the bottom of a deep well in the town of Syene (now Aswan) in Upper Egypt; that is, on that day the sun was exactly overhead at noon. But in Alexandria, due north of Syene, the sun at that moment was one-fiftieth of a full circle (i.e., 7.2°) from the zenith, as measured by the shadow of a vertical rod (fig. 23). Eratosthenes assumed that the sun is so far away from the earth that its rays reach us practically parallel; hence the diffference in the sun's elevation as seen from the two locations must be due to the sphericity of the earth. Since the distance between Alexandria and Syene was 5,000 stadia (as measured by the time it took the king's messengers to run between the two cities), the circumference of the earth must be fifty times this distance, or 250,000 stadia.

Unfortunately, the exact length of the *stadium*, the geographical distance unit in the Greek era, is not known; estimates vary from 607 to 738 feet, the smaller figure referring to the Roman stadium of later use. The circumference of the earth as found

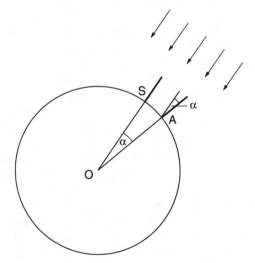

FIG. 23. Eratosthenes's measurement of the earth's circumference.

by Eratosthenes is therefore between 29,000 and 35,000 miles, whereas the correct values are 24,818 miles for the polar circumference and 24,902 miles for the equatorial.[6] Eratosthenes thus came remarkably close, and in doing so he used the science of geometry in its literal sense: to measure the earth.

✧ ✧ ✧

Ferdinand Magellan's historic circumnavigation of the globe (1519–1522) provided the first direct proof that the earth is roughly spherical. But soon thereafter scientists began to suspect that it might actually be flattened; the question was, was it flattened at the poles (an *oblate spheroid*) or at the equator (a *prolate spheroid*)? This was not merely an academic issue: with the age of exploration well under way, it became crucial for navigators to be able to determine their position at sea—their longitude and latitude—with sufficient accuracy. This in turn depended on knowing the length of a degree of latitude as measured along a meridian (a "meridional degree"). Had the earth been a perfect sphere, a degree would have the same length everywhere, regardless of the latitude itself. But if the earth was oblate, the length of a degree would increase slightly toward the poles, while if it was prolate, the length would decrease. Determining the exact shape of the earth—and more generally, of any curved surface—evolved into the science of *geodesy*. Some of the greatest mathematicians of the eighteenth and nineteenth centuries tackled this problem, among them Newton, Euler and Gauss.

The first step in a geodetic survey is to select a baseline of known length, and then measure the angles between this line and the lines of sight from its endpoints to a distant landmark. For relatively small regions—a town or a county—one may ignore the earth's curvature and regard the region as planar; one can then compute the distance from either endpoint to the landmark, using the Law of Sines (the *ASA* case). These distances can now be used as new baselines and the process repeated, until the entire region is covered by a network of triangles. This process is known as *triangulation*; it provides the skeleton on which the topographical details of the land—hills, rivers, lakes, towns, and roads—are later superimposed to form a complete map.[7]

The method of triangulation was suggested as early as 1533 by the Dutch mathematician Gemma Frisius (1508–1555).[8] It was first carried out on a large scale by another Dutchman, Willebrord van Roijen Snell (1581–1626),[9] who in 1615 surveyed a stretch of 80 miles in Holland, using a grid of 33 triangles. But it was in France that the first comprehensive, government-sponsored triangulation effort was undertaken in 1668 under the direction of the Abbé Jean Picard (1620–1682), one of the founders of the Paris Observatory. As his baseline he chose a 7 mile stretch along the road from Paris to Fontainbleau; from this baseline the survey would eventually cover all of France. To improve the accuracy of his measurements, Picard used a new type of quadrant (an instrument for measuring vertical angles) in which a telescope with cross hairs replaced the two pinholes for sighting. Combining his terrestrial measurements with an astronomical determination of the latitudes of the endpoints of his baseline, Picard arrived at the value of 68.65 miles for the degree at the latitude of Paris. He then extended his survey to the French coastline, which resulted in an unwelcome discovery: the west coast of the country had to be shifted $1\frac{1}{2}^{\circ}$ eastward relative to the prime meridian through Paris, causing the monarch Louis XIV to exclaim: "Your journey has cost me a major portion of my realm!"[10]

After Picard's death in 1682 the survey was continued for another century by four generations of a remarkable family of astronomers, the Cassini family. Giovanni Dominico Cassini (1625–1712) was born in Italy and taught at the University of Bologna, but in 1668, as a result of Picard's persistent efforts, he left his post and became head of the newly founded Paris Observatory. Changing his name to Jean Dominique, he made significant contributions to astronomy: a determination of the rotation periods of Mars and Jupiter, the first study of the zodiacal light (a faint glow that stretches up the eastern sky before sunrise and the western sky after sunset), the discovery of four of Saturn's satellites and a dark gap in Saturn's rings (known as the Cassini Division), and a measurement of the parallax of Mars in 1672 from which he was able to calculate—using Kepler's laws of planetary motion—the distance from the earth to the sun as 87 million miles, the first determination to come close to the actual value of about 93.5 million miles. Amazingly, he was also one of the last professional astronomers to oppose Copernicus's heliocentric system, and he remained convinced that the earth was a prolate spheroid, despite mounting evidence to the contrary.[11]

In later life Cassini increasingly devoted himself to geodesy and cartography, using his astronomical expertise to the fullest. In 1679 he conceived a new world map, the *planisphère terrestre*, using a projection on which all directions and distances from the north pole are shown correctly; this is known as the *azimuthal equidistance projection* (see chapter 10). Cassini's huge map, 24 feet in diameter, was drawn on the third floor of the Paris Observatory; it became a model for future cartographers and was reproduced and published in 1696.

But Cassini did not rest on his laurels. Now seventy, he resumed the survey of France with renewed vigor, assisted by his son Jacques (1677–1756). Their goal: to extend the survey south to the Pyrenees and eventually cover all of Europe with a network of triangles. As a by-product, they hoped to find out whether the earth was prolate or oblate.

The elder Cassini died in 1712 at the age of eighty-seven. His son now realized that the issue could only be resolved by comparing the length of a degree at widely separated latitudes, and he suggested that expeditions be sent to the equator and the arctic region to settle the question once and for all. At stake was not only a theoretical interest in the shape of the earth; the prestige of France itself was in the balance. Newton had predicted that the earth is flattened at the poles, basing his argument on the interaction between the gravitational pull of the earth on itself and the centrifugal forces due to its spinning on its axis. In France, however, Newton's ideas on gravitation—especially his notion of "action at a distance"—were rejected in favor of Descartes' theory of vortices, which held that gravitational attraction is caused by huge vortices swirling around in a fluid that permeates all space. "The figure of the earth had become a *cause célèbre*, the most debated scientific issue of the day, with French and British national pride at stake."[12]

The evidence, though indirect, tended to support Newton. For one, the planet Jupiter as seen through even a small telescope shows considerable flattening at the poles; and here on earth, measurements of the acceleration due to gravity, as obtained from the period of a swinging pendulum, showed a slightly smaller value at the equator than at the poles, indicating that the equator is farther away from the center of the earth than the poles.

Following Jacques Cassini's suggestion and with the blessing of the new monarch, Louis XV, the Académie Royale de Sciences—the French equivalent of the Royal Society in England—in 1734 authorized two expeditions, one to be sent to Lapland on the Swedish-Finnish border, the other to Peru, close to the equator. Their mission: carry out a full-scale triangulation of their respective regions and determine the length of the degree at each location.

The first expedition was headed by Pierre Louis Moreau de Maupertuis (1698–1759), who began his career in the French army and later became a mathematician and physicist (he was the first to formulate the principle of least action and later used it to "prove" the existence of God). As a lone supporter and admirer of Newton on the Continent, he was only too eager to join an endeavor that, he hoped, would prove his master right. Going with him was another French mathematician, Alexis Claude Clairaut (1713–1765), who as a young prodigy had studied calculus when he was ten and published his first book at eighteen (the differential equation $xy' - y = f(y')$, where f is a given function of the derivative y', is named after him). The Peru expedition was headed by a geographer, Charles Marie de La Condamine (1701–1774), and it too included a mathematician, Pierre Bouger (1698-1758). The participation of so many first-rank mathematicians in field trips to remote countries was in line with a long French tradition of producing eminent scientists who also served their country in the military and civil service. We will meet some more of them in chapter 15.

The two expeditions encountered numerous hardships. The Lapland team braved blinding snowstorms and had to force its way across frozen marshes that turned into mud when spring arrived; during summer their biggest enemy was the biting mosquitoes. The Peru team fared even worse: not only did they combat altitude sickness in the high Andes, but disease and a series of accidents resulted in the death of several members. To top it all, bitter dissent broke out among the expedition leaders, who returned home separately. Nevertheless the two expeditions accomplished their tasks: they found the length of a degree to be 69.04 miles in Lapland and 68.32 miles in Peru (fig. 24). Combined with Picard's value of 68.65 miles at Paris, their results proved beyond doubt that the earth is an oblate spheroid. Newton was proven right—again.

And now the Cassinis returned to the scene. While the two expeditions were doing their work abroad, Jacques Cassini and his son César François (1714–1784) completed the triangulation of France, using a network of eighteen baselines and four hundred triangles. It now remained to transform the grid into an actual map, and this task was completed by Jean Dominique Cassini IV (1748–1845), great-grandson of the dynasty's founder. His huge map, 12 by 12 yards, was published in 182 sheets on a scale of 1:86,400 and showed not only topographical features but also the location of castles, windmills, vineyards, and—this being the time of the French Revolution—guillotines. The fourth Cassini received much praise for his work—and then was arrested and tried by a revolutionary tribunal, barely saving his life. His rep-

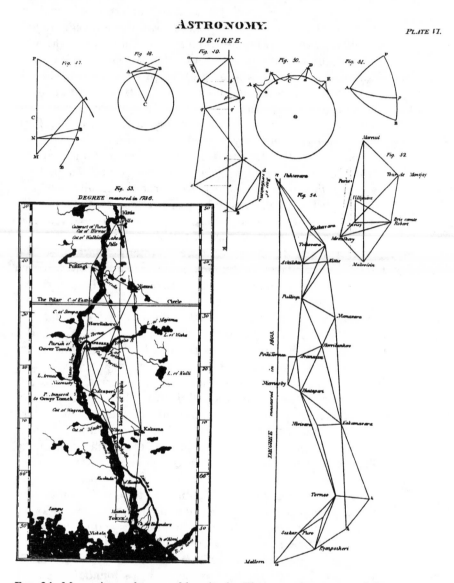

FIG. 24. Measuring a degree of longitude. The map shows part of the triangulation network set up by Maupertuis's expedition to Lapland. From a 1798 engraving (the author's collection).

utation was eventually restored by Napoleon Bonaparte, and he died in 1845 at the age of ninety-seven.

✧ ✧ ✧

France's geodetic lead was now followed by the rest of Europe, and by the mid-nineteenth century most of the continent was thoroughly triangulated and mapped. The task then moved overseas to India, the crown jewel of the British Empire, were a huge triangulation project, known as the Great Trigonometrical Survey, was undertaken from 1800 to 1913. Sponsored by the East India Company, the vast commercial eneterprise that virtually ruled the country from its offices in London, the survey began near Madras on the southwest coast of the Bay of Bengal and eventually reached the Himalayas in the far north.

Captain William Lambton (1753–1823), who headed the survey from 1802 until his death, was determined to achieve his goal with unprecedented accuracy. His huge theodolite, weighing half a ton, was built in London according to his specific instructions and shipped to India, being intercepted en route by a French frigate. On one occasion this monstrous instrument was hoisted to the top of the Great Temple of Tanjore (Thanjavur) so as to give the surveyors a clear view of the terrain. On the way up a rope snapped and the instrument fell to the ground and broke. Undaunted, Lambton secluded himself in his tent and for the next six weeks repaired the instrument himself.

In 1806 Lambton set out to achieve a goal even greater than the survey of India: to determine the figure of the earth. To this end he ran a line along the 78° meridian from near Cape Comorin at the southern tip of the subcontinent to the Kashmir region in the north, a distance of some 1,800 miles. His men encountered numerous perils: the intense heat of central India, thick vegetation in which tigers were roaming, the ever present threat of malaria, and angry locals who were convinced the surveyors were spying on their wives.

After Lambton's death, the survey continued under his assistant George (later Sir George) Everest (1790–1866), who would become Surveyor General of India. Everest maintained and even exceeded the high standards of his predecessor. To make up for the absence of natural landmarks in the vast plains of central India, he built a series of towers visible from far away, many of which still stand today. And to avoid the heat and haze in the country's interior, he ordered his surveyors to work at night, relying on bonfires lit from the top of his towers as

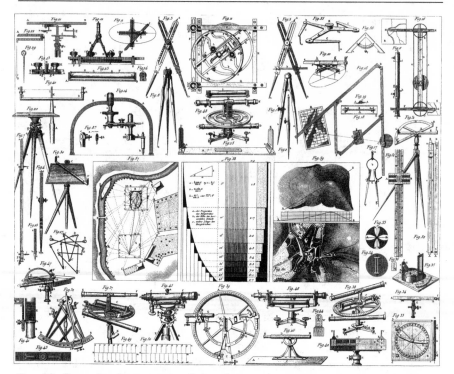

FIG. 25. Surveying instruments of the nineteenth century.

signals. During daytime he used a heliograph—a specially designed mirror whose reflected sunlight could be seen from 50
miles away. Everest's meticulous care for detail paid off: when
the survey reached the foothills of the Himalayas, his actual position differed from that computed from his triangulation by 7
inches over a distance of 500 miles!

During his survey Everest made a discovery that is still being debated today: he found that the great mass of the Him-
layan mountains deflected the direction of the plumb line. These
gravity anomalies were the first indication of what is now called
"mascons" (mass concentrations, the term having first been applied to the moon), whose exact nature and distribution is being
mapped today by satellites.

After Everest's retirement in 1843 the survey continued under
the direction of his assistant, Captain Andrew Waugh. And now
the power of trigonometry to measure faraway objects reached
its high point—literally. Legend has it that one day in 1852 the
survey's chief computer, Radhanath Sikdar, himself a mathematician, burst into Waugh's office exclaiming, "Sir, I have discovered the highest mountain in the world." The official an-

nouncement was delayed until 1856 so as to check and recheck the height of Peak XV, as Mount Everest was temporarily named (it is also known by its Tibetan name Chomolungma, "goddess mother of the world"). Averaging several readings taken at a distance of some 100 miles, the mountain's height turned out exactly 29,000 feet; but fearing that such a round figure might appear as if it had been made up, the surveyor arbitrarily added two feet, and until 1954 the official height of the world's highest peak remained 29,002 feet above sea level. It is now put at 29,028 feet.[13]

✧   ✧   ✧

While the French were busy triangulating Europe and the British surveyed their empire, Friedrich Wilhelm Bessel (1784–1846) in Germany was set on triangulating the heavens. Beginning his career as an accountant, he taught himself mathematics and astronomy and at the age of twenty recalculated the orbit of Halley's comet, taking into account the gravitational perturbations exerted on it by the planets Jupiter and Saturn. Bessel's achievements brought him to the attention of the leading German astronomer at the time, Heinrich Olbers, who secured him a position at the observatory of Lilienthal. His reputation as a skilled observational astronomer as well as a theoretician of the first rank led to his appointment in 1809 as director of the Prussian Royal Observatory at Königsberg (now Kaliningrad in Russia), a position he kept until his death.[14]

By 1800 the size of the then-known solar system was fairly well established (though the planets Neptune and Pluto had yet to be discovered), but the dimensions of the universe beyond was a different matter: no one had the faintest idea how far were the fixed stars. The method of parallax, so successful in determining the distances of solar system objects, had thus far failed when applied to the fixed stars: no star showed any measurable shift in position even when using the largest baseline available to us—the diameter of earth's orbit around the sun. In fact, the absence of parallax was taken by the Greeks as the strongest evidence for their picture of a universe in which a motionless earth was permanently fixed at the center. Copernicus interpreted the facts differently: to him the absence of any discernible parallax indicated that the stars are so far away from us that any apparent shift in their position due to earth's motion around the sun would be far too small to be detected by our eyes.

When the telescope was invented in 1609 it became theoretically possible to look for the parallax of some nearby stars, but all such attempts have failed. One reason was that astronomers were looking only at the brightest stars in the sky, assuming that their brightness indicated that they were also the nearest. This would be true if all stars had the same intrinsic brightness—the same light output—like street lights lining up an avenue. But by 1800 astronomers knew that stars differ vastly in their intrinsic brightness, and consequently their *apparent* brightness could not be used as an effective yardstick in estimating their relative distance. Instead, the search turned to stars with a large *proper motion*—the actual motion of a star relative to the distant sky (as opposed to apparent motion, which is merely due to the observer's own motion). It was correctly assumed that a large proper motion would be an indication that the star is relatively close by.

Soon a candidate was found—the star 61 Cygni in the constellation Cygnus, the Swan. This fifth-magnitude star is barely visible to the naked eye but was known to have a considerable proper motion—5.2 arc seconds per year, or about one moon diameter every 350 years. Bessel now directed all his efforts at this star. After 18 months of intense observation, he announced in 1838 that 61 Cygni showed a parallax of 0.314 arc seconds (by comparsion, the moon's apparent diameter is about half a *degree* of arc, or 1,800 arc seconds). This minute figure was enough for Bessel to find the star's distance, using the simplest of trigonometric calculations. Because this event marks a milestone in the history of astronomy, we give the details here.

In figure 26 let S represent the sun, E_1 and E_2 the earth at opposite positions in its orbit around the sun, and T the star in question. By astronomical convention, the annual parallax is defined as *half* the angular shift in the star's position due to earth's motion around the sun, that is, the angle $\alpha = \angle E_1 TS$ in the right triangle $E_1 ST$. Denoting the distance to the star by d and the radius of earth's orbit by r, we have $\sin \alpha = r/d$, or

$$d = \frac{r}{\sin \alpha}.$$

Substituting the values $r = 150{,}000{,}000$ km $= 1.5 \times 10^8$ km and $\alpha = 0.314'' = (0.314/3{,}600)°$, we get $d = 9.85 \times 10^{13}$ km. Now, stellar distances are usually expressed in light years, so we must divide this figure by the speed of light, 3×10^5 km/sec, times the number of seconds in a year, $3{,}600 \times 24 \times 365$. This gives us

$d = 10.1$ light years.

Thus the dimensions of the universe beyond our solar system became known. The parallax of 61 Cygni has since been refined

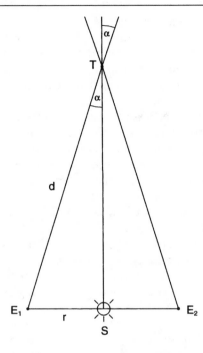

FIG. 26. Stellar parallax.

to 0.294″, resulting in a distance of 11.1 light years. Soon other stellar parallaxes were successfully measured, among them that of Alpha Centauri, which at 4.3 light years is our closest celestial neighbor beyond the sun.[15] The method has been applied to stars up to about a hundred light years away, but for greater distances its accuracy quickly diminishes. Fortunately other methods, based on a star's physical characteristics, have been developed in the strive to establish a reliable stellar distance scale.

In 1844 Bessel made a second epochal discovery: he directed his telescope to Sirius, the brightest star in the sky, and found that its proper motion shows a minute wavelike pattern; this he correctly attributed to the gravitational influence of an invisible companion that revolves around Sirius. This companion, Sirius B, was discovered in 1862 by the American telescope maker Alvan Graham Clark (1832–1897).

Gravitational perturbations have occupied Bessel's mind through much of his life. This subject presented one of the most difficult problems in celestial mechanics, and to deal with it he introduced a certain class of functions known since as the *Bessel functions.* These are solutions of the differential equation $x^2 y'' + xy' + (x^2 - n^2)y = 0$, where $n \geq 0$ is a constant (not necessarily an integer). The nature of the solutions greatly depends on n: for $n = 1/2, 3/2, 5/2, \ldots$, they can be written in closed form in terms of the functions x, $\sin x$ and $\cos x$; otherwise they can only be expressed as infinite series and are therefore considered "nonelementary" or "higher" functions. The Bessel equation shows up in many applications; for example, the vibrations of a circular membrane—such as a drum head—are governed by Bessel's equation with $n = 0$.[16]

Toward the end of his life Bessel turned again to the problem of gravitational perturbations. One of the outstanding astronomical riddles at the time was the anomalies in the motion of the planet Uranus: all attempts to explain these anomalies as being caused by known planets—in particular, Jupiter and Saturn—had failed. Bessel correctly attributed them to the existence of an unknown transuranian planet, but he died a few months before this planet, Neptune, was discovered.

Bessel was one of the last great scientists who was equally at home in theory as in experiment (in this case, observational techniques). Mathematicians will remember him for the Bessel functions, but his crowning achievement was in giving us the first concrete evidence of how vast interstellar space really is. With him, the attention of astronomers began to shift from the solar system to the universe beyond.

NOTES AND SOURCES

1. See Sir Thomas. L. Heath, *Aristarchus of Samos: the Ancient Copernicus* (1913; rpt. New York: Dover, 1981), and *Greek Astronomy* (1932; rpt. New York: Dover, 1991).

2. This follows from the fact that for $0° < x < 90°$, the graph of $(\sin x)/x$ is decreasing, while that of $(\tan x)/x$ is increasing; that is, $(\sin \alpha)/\alpha < (\sin \beta)/\beta$ and $(\tan \alpha)/\alpha > (\tan \beta)/\beta$.

3. In 1995 I joined a group of astronomers on an expedition to India to watch the total solar eclipse of October 24. From our sight close to the center line of the moon's shadow, totality lasted a mere 41 seconds.

4. Quoted from Bryan Brewer, *Eclipse* (Seattle: Earth View, 1978), p. 31.

5. See Albert van Helden, *Measuring the Universe: Cosmic Dimensions from Aristarchus to Halley* (Chicago: University of Chicago Press, 1985), p. 11. See also Toomer's article on Hipparchus in the *DSB*.

6. Much scholarly debate has been devoted to the length of the stadium. Some sources give it as equal to one-tenth of a mile, or 528 feet, which would lead to a circumference of 25,000 miles. It seems, though, that this length of the stadium has been "fixed" in retrospect so as to make the circumference come close to the modern value. To quote B. L. van der Waerden in *Science Awakening* (New York: John Wiley, 1963), p. 230: "Since we do not know the exact length of a stadium, we

can say little more than that the order of magnitude [of the earth's circumference] is about right." See also David Eugene Smith, *History of Mathematics* (1925; rpt. New York: Dover, 1958), vol. 2, p. 641.

7. The material for the narrative that follows is based on the following sources: Lloyd A. Brown, *The Story of Maps* (1949; rpt. New York: Dover, 1979); John Noble Wilford, *The Mapmakers* (New York: Alfred A. Knopf, 1981); and Simon Berthon and Andrew Robinson, *The Shape of the World* (Chicago: Rand McNally, 1991).

8. His real name was Gemma Regnier, but he became known as Gemma Frisius after his place of birth, Friesland. In 1541 he became professor of medicine at the University of Louvaine. His book on arithmetic (Antwerp, 1540) was highly popular and went through no fewer than sixty editions. He also wrote on geography and astronomy and suggested the method of determining longitude from the difference in local time between two places. His son Cornelius Gemma Frisius (1535–1577) continued his father's work and served as professor of medicine at the same university.

9. He was professor of mathematics at Leyden, where he succeeded his father. He worked on astronomy, physics, and spherical trigonometry and is best known for his law of refraction in optics.

10. It was not until 1913 that France recognized the meridian through Greenwich as the prime (zero) meridian, in exchange for England "recognizing" the metric system.

11. The Cassini spacecraft launched by NASA in October 1997 on a seven-year voyage to Saturn is named after him.

12. Quoted from Berthon and Robinson, *Shape of the World*, p. 101.

13. Surprisingly, in his autobiography, *Nothing Venture, Nothing Win* (New York: Coward, McCann & Geoghegan, 1975), Sir Edmund Hillary, who with Sherpa Tenzing Norgay was the first to climb Mount Everest in 1953, still gives the mountain's height as 29,002 feet—more than twenty years after it had officially been changed. In 1994 a Chinese surveying team, aided by a global satellite positioning system, put the figure at 29,023 feet.

14. See Walter Fricke's article on Bessel in the *DSB*.

15. Actually, Alpha Centauri is a triple star system, whose faintest companion, Proxima Centauri (discovered in 1915), is at present 4.2 light years away. 61 Cygni is now ranked nineteenth in the order of distance from the sun. See the article, "Our Nearest Celestial Neighbors," by Joshua Roth and Roger W. Sinnott, *Sky & Telescope*, October 1996, pp. 32–34.

16. For $n = 0$ and 1 the Bessel functions—denoted by $J_0(x)$ and $J_1(x)$—exhibit certain similarities to $\cos x$ and $\sin x$, respectively; for example, $J_0(0) = 1$ and $J_1(0) = 0$, and both functions have an oscillating graph. However, their amplitudes diminish as x increases, and their zeros are not equally spaced along the x-axis, which is why the sound of a drum is different from that of a violin (see chapter 15). For details, see any book on ordinary differential equations.

Abraham De Moivre

Abraham De Moivre was born in Vitry in the province of Champagne, France, on May 26, 1667, to a Protestant family. He showed an early interest in mathematics and studied it—secretly—at the various religious schools he was attending. In 1685 Louis XIV revoked the Edict of Nantes—a decree issued in 1598 granting religious freedom to French Protestants—and a period of repression followed. By one account De Moivre was imprisoned for two years before leaving for London, where he would spend the rest of his life. He studied mathematics on his own and became very proficient in it. By sheer luck he happened to be at the house of the Earl of Devonshire, where he worked as a tutor, at the very moment when Isaac Newton stepped out with a copy of the *Principia*, his great work on the theory of gravitation. De Moivre took up the book, studied it on his own, and found it far more demanding than he had expected (it is a difficult text even for a modern reader). But by assiduous study—he used to tear pages out of the huge volume so he could study them between his tutoring sessions—he not only mastered the work but became an expert on it, so much so that Newton, in later years, would refer to De Moivre questions addressed to himself, saying, "Go to Mr. De Moivre; he knows these things better than I do."

In 1692 he met Edmond Halley (of comet fame), who was so impressed by his mathematical ability that he communicated to the Royal Society De Moivre's first paper, on Newton's "method of fluxions" (i.e., the differential calculus). Through Halley, De Moivre became a member of Newton's circle of friends that also included John Wallis and Roger Cotes. In 1697 he was elected to the Royal Society and in 1712 was appointed as member of the Society's commission to settle the bitter priority dispute between Newton and Leibniz over the invention of the calculus. He was also elected to the academies of Paris and Berlin.

Despite these successes, De Moivre was unable to secure himself a university position—his French origin was one reason—and even Leibniz's attempts on his behalf were unsuccessful. He made a meager living as a tutor of mathematics, and for the rest of his life would lament having to waste his time walking between the homes of his students. His free time was spent in the coffeehouses and taverns on St. Martin's Lane in London, where he answered all kinds of mathematical questions addressed to him by rich patrons, especially about their chances of winning in gambling.

When he grew old he became lethargic and needed longer sleeping hours. According to one account, he declared that beginning on a certain day he would need twenty more minutes of sleep on each subsequent day. On the seventy-third day—November 27, 1754—when the additional sleeping time accumulated to 24 hours, he died; the official cause was recorded as "somnolence" (sleepiness). He was eighty-seven years old, joining a long line of distinguished English mathematicians who lived well past their eighties: William Oughtred, who died in 1660 at the age of 86, John Wallis (d. 1703 at 87), Isaac Newton (d. 1727 at 85), Edmond Halley (d. 1742 at 86), and in our time, Alfred North Whitehead (d. 1947 at 86) and Bertrand Russell, who died in 1970 at 98. The poet Alexander Pope paid him tribute in *An Essay on Man*:

Who made the spider parallels design,
Sure as Demoivre, without rule or line?

❖ ❖ ❖

De Moivre's mathematical work covered mainly two areas: the theory of probability, and algebra and trigonometry (considered as a unified field). In probability he extended the work of his predecessors, particularly Christiaan Huygens and several members of the Bernoulli family. A generalization of a problem first posed by Huygens is known as *De Moivre's problem*: Given n dice, each having f faces, find the probability of throwing any given number of points.[1] His many investigations in this field appeared in his work *The Doctrine of Chances: or, a Method of Calculating the Probability of Events in Play* (London, 1718, with expanded editions in 1738 and 1756); it contains numerous problems about throwing dice, drawing balls of different colors from a bag, and questions related to life annuities. Here also is stated (though he was not the first to discover it) the law for finding the probability of a compound event. A second work, *A Treatise of Annuities upon Lives* (London, 1725 and 1743), deals with the analysis of mortality statistics (which Halley had begun some years earlier), the division of annuities among several heirs, and other questions of interest to financial institutes and insurance companies.

In the theory of probability one constantly encounters the expression $n!$ (read "n factorial"), defined as $1 \cdot 2 \cdot 3 \cdot \cdots \cdot n$. The value of $n!$ grows very rapidly with increasing n; for example, $10! = 3,628,800$ while $20! = 2,432,902,008,176,640,000$. To find $n!$ one must first find $(n - 1)!$, which in turn requires finding $(n - 2)!$ and so on, making a direct calculation of $n!$ for large n extremely time consuming. It is therefore desirable to have an approximation formula that would estimate $n!$ for large n by a single calculation. In a paper written in 1733 and presented privately to some friends, De Moivre developed the formula

$$n! \approx cn^{n+1/2}e^{-n},$$

where c is a constant and e the base of natural logarithms.[2] He was unable, however, to determine the numerical value of this constant; this task befell a Scot, James Stirling (1692–1770), who found that $c = \sqrt{2\pi}$. Stirling's formula, as it is known today, is thus as much due to De Moivre; it is usually written in the form

$$n! \approx \sqrt{2\pi n}\, n^n e^{-n}.$$

As an example, for $n = 20$ the formula gives $2.422, 786, 847 \times 10^{18}$, compared to the correct, rounded value $2.432, 902, 008 \times 10^{18}$.

De Moivre's third major work, *Miscellanea Analytica* (London, 1730), deals, in addition to probability, with algebra and analytic trigonometry. A major problem at the time was how to factor a polynomial such as $x^{2n} + px^n + 1$ into quadratic factors. This problem arose in connection with Cotes' work on the decomposition of rational expressions into partial fractions (then known as "recurring series"). De Moivre completed Cotes's work, left unfinished by the latter's early death (see p. 182). Among his many results we find the following formula, sometimes known as "Cotes' property of the circle":

$$x^{2n} + 1 = [x^2 - 2x \cos \pi/2n + 1][x^2 - 2x \cos 3\pi/2n + 1]$$
$$\cdots [x^2 - 2x \cos (2n - 1)\pi/2n + 1].$$

To obtain this factorization, we only need to find (using De Moivre's theorem) the $2n$ different roots of the equation $x^{2n} + 1 = 0$, that is, the $2n$ complex values of $\sqrt[2n]{-1}$, and then multiply the corresponding linear factors in conjugate pairs. The fact that trigonometric expressions appear in the factorization of a purely algebraic expression such as $x^{2n} + 1$ amazes any student who encounters such a formula for the first time; in De Moivre's time it amazed even professional mathematicians.

✧   ✧   ✧

De Moivre's famous theorem,

$$(\cos \phi + i \sin \phi)^n = \cos n\phi + i \sin n\phi,$$

was suggested by him in 1722 but was never explicitly stated in his work; that he knew it, however, is clear from the related formula

$$\cos \phi = \tfrac{1}{2}(\cos n\phi + i \sin n\phi)^{1/n} + \tfrac{1}{2}(\cos n\phi - i \sin n\phi)^{1/n},$$

which he had already found in 1707 (De Moivre derived it for positive integral values of n; Euler in 1749 proved it for any real n).[3] He frequently used it in *Miscellanea Analytica* and in numerous papers he published in the *Philosophical Transactions*, the official journal of the Royal Society. For example, in a paper published in 1739 he showed how to extract the roots of any binomial of the form $a + \sqrt{b}$ or $a + \sqrt{-b}$ (he calls the latter an "impossible binomial"). As a specific example he shows how to find the three cube roots of $81 + \sqrt{-2{,}700}$ (in modern notation, $81 + (30\sqrt{3})i$). The discussion is verbal rather than symbolic, but it is precisely the method we find today in any trigonometry textbook: write $81 + (30\sqrt{3})i$ in polar form as $r(\cos \theta + i \sin \theta)$, where $r = \sqrt{[81^2 + (30\sqrt{3})^2]} = \sqrt{9{,}261} = 21\sqrt{21}$ and $\theta = \tan^{-1}(30\sqrt{3})/81 = \tan^{-1}(10\sqrt{3})/27 = 32.68°$. Then compute the expression $(\sqrt[3]{r})[\cos(\theta + 360°k)/3 + i \sin(\theta + 360°k)/3]$ for $k = 0, 1,$ and 2. We have $\sqrt[3]{(21\sqrt{21})} = (21^{3/2})^{1/3} = 21^{1/2} = \sqrt{21}$ and $\theta/3 = 10.89°$, so the roots are $\sqrt{21}$ cis $(10.89° + 120°k)$, where "cis" stands for $\cos + i \sin$. Using a table or a calculator to find the sines and cosines, we get the three required roots: $(9 + (\sqrt{3})i)/2$, $-3 + (2\sqrt{3})i$, and $(-3 - (5\sqrt{3})i)/2$. De Moivre adds:

There have been several authors, and among them the eminent Wallis, who have thought that those cubic equations which are referred to the circle, may be solved by the extraction of the cube root of an imaginary quantity, and of $81 + \sqrt{-2{,}700}$, without regard to the table of sines, but this is a mere fiction and a begging of the question. For on attempting it, the result always recurs back again to the same question as that first proposed. And the thing cannot be done directly, without the help of the table of sines, especially when the roots are irrational, as has been observed by many others.[4]

De Moivre must have surely wondered why the three roots come out as "nice" irrational complex numbers, even though the angle θ is not any "special" angle like $15°, 30°$, or $45°$. He says that it is a "fiction" (i.e., impossible) to find the cubic root of a complex number without a table of sines; and to avoid any misunderstanding, he repeats this statement again toward the end: "And the thing cannot be done directly, without the help of the table of sines, especially when the roots are irrational." He is of course right in the general case: to find the three cubic roots of a complex number $z = x + iy$, we must express it in polar form, $z = r$ cis θ, where $r = \sqrt{(x^2 + y^2)}$ and $\theta = \tan^{-1} y/x$; next we compute $\sqrt[3]{r}$ and $\theta/3$, then—using a table of sines— we find $\cos \theta/3$ and $\sin \theta/3$, and finally $(\sqrt[3]{r})$ cis $(\theta/3 + 120°k)$ for $k = 0, 1$, and 2. Ironically, however, the very example De Moivre brings to illustrate the procedure could be solved without recourse to a table! Let us see how.

We wish to find the three cubic roots of $z = x + iy = 81 + (30\sqrt{3})i = r$ cis θ, where $r = 21\sqrt{21}$ and $\tan \theta = (10\sqrt{3})/27$. From this last equation (or by directly computing x/r) we find that $\cos \theta = 81/(21\sqrt{21}) = (9\sqrt{21})/49$. We now use the identity $\cos \theta = 4\cos^3 \theta/3 - 3 \cos \theta/3$ to find the value of $\cos \theta/3$; letting $x = \cos \theta/3$, we have

$$(9\sqrt{21})/49 = 4x^3 - 3x \tag{1}$$

or

$$196x^3 - 147x - 9\sqrt{21} = 0. \tag{2}$$

The substitution $y = x/(9\sqrt{21})$ reduces this equation to

$$333,396\, y^3 - 147y - 1 = 0. \tag{3}$$

This new equation is radical-free, but its leading coefficient looks hopelessly large. It so happens, however, that 147 is divisible by 21 and 333,396 is divisible by 21^3. Writing $z = 21y$, the equation becomes

$$36z^3 - 7z - 1 = 0, \tag{4}$$

a pretty simple equation whose three roots are $1/2, -1/3$ and $-1/6$—all rational numbers! Substituting back, we have $y = z/21 = 1/42, -1/63$ and $-1/126$, and finally $x = \cos \theta/3 = (9\sqrt{21})y = (3\sqrt{21})/14, -(\sqrt{21})/7$, and $-(\sqrt{21})/14$. For each of these values we now find $\sin \theta/3$ from the identity $\sin \theta/3 = \pm\sqrt{(1 - \cos^2 \theta/3)}$; we get $\sin \theta/3 = (\sqrt{7})/14, (2\sqrt{7})/7$, and $-(5\sqrt{7})/14$ (the last is negative because the corresponding point is in Quadrant III of the complex plane). We also have $\sqrt[3]{r} = (21\sqrt{21})^{1/3} = \sqrt{21}$. The three required roots are thus

$$\sqrt{21}\left[3\sqrt{21}/14 + (\sqrt{7}/14)i\right] = \tfrac{1}{2}(9 + \sqrt{3}i),$$
$$\sqrt{21}\left[-\sqrt{21}/7 + (2\sqrt{7}/7)i\right] = -3 + \sqrt{3}i,$$

and

$$\sqrt{21}\left[-\sqrt{21}/14 - (5\sqrt{7}/14)i\right] = \tfrac{1}{2}(-3 - 5\sqrt{3}i);$$

they are shown in figure 27.

Of course, the "natural" way of handling this problem would be to solve equation (1) directly, using a formula named after the Italian Girolamo Cardano (also known as Jerome Cardan, 1501–1576) but actually developed independently by two other Italians, Scipione del Ferro (ca. 1465–1526) and Nicolo Tartaglia (ca. 1506–1557).[5] Cardano's formula is analogous to the familiar quadratic formula for solving equations of the second degree but is considerably more complicated; it is based on the fact that any cubic equation in normal form, $y^3 + ay^2 + by + c = 0$ (where the leading coefficient $= 1$) can be brought to the *reduced* form $x^3 + px + q = 0$ (with no quadratic term) by the substitution $y = x - a/3$, where $p = b - a^2/3$ and $q = 2a^3/27 - ab/3 + c$. Since equation (1) is already free of a quadratic term, we only need to divide it by its leading coefficient, getting $x^3 + px + q = 0$, where $p = -3/4$ and $q = -(9\sqrt{21})/196$. Cardano's formula now requires one to compute the quantities $P = \sqrt[3]{-q/2 + \sqrt{q^2/4 + p^3/27}}$ and $Q = \sqrt[3]{-q/2 - \sqrt{q^2/4 + p^3/27}}$. Substituting the values of p

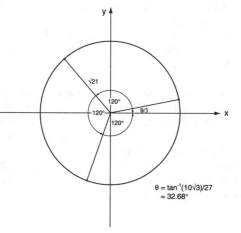

$$\theta = \tan^{-1}(10\sqrt{3})/27$$
$$= 32.68°$$

FIG. 27. The three cube roots of $81 + \sqrt{-2700}$.

and q into these expressions we get, after considerable simplification, $P, Q = \frac{1}{14} \sqrt[3]{-63\sqrt{21} \pm (70\sqrt{7})i}$. So we must now find the cubic root of the complex numbers $-63\sqrt{21} \pm (70\sqrt{7})i$, and to do so we must express them in polar form R cis ϕ. We have $R = \sqrt{[(-63\sqrt{21})^2 + (70\sqrt{7})^2]} = 343$ and $\phi = \pm\tan^{-1}(70\sqrt{7})/(63\sqrt{21}) = \pm\tan^{-1}(10\sqrt{3})/27$—the very same angle we had in the first place! This is what De Moivre meant in his enigmatic statement, "For on attempting it, the result always recurs back again to the same question as that first proposed."

Could a mathematician of De Moivre's caliber have overlooked the fact that his own example could be solved without using a table of sines? Apparently he did. Even Einstein once ignored the possibility that a denominator in one of his equations might be zero. This was in 1917, when he applied his general theory of relativity to cosmological questions. A young Russian astronomer, Aleksandr Friedmann, noticed this seemingly benign oversight and concluded that the particular case Einstein had ignored implied, no less, that the universe might be expanding![6]

NOTES AND SOURCES

1. Florian Cajori, *A History of Mathematics* (1893; 2d ed. New York: Macmillan, 1919), p. 230.

2. This paper also gives the first statement of the formula for the normal distribution. See David Eugene Smith, *A Source Book in Mathematics* (1929; rpt. New York: Dover, 1959), pp. 566–568.

3. From this relation and its companion,

$$i\sin\phi = \frac{1}{2}(\cos n\phi + i\sin n\phi)^{1/n} - \frac{1}{2}(\cos n\phi - i\sin n\phi)^{1/n}$$

we get, upon adding, $\cos\phi + i\sin\phi = (\cos n\phi + i\sin n\phi)^{1/n}$, from which De Moivre's theorem follows immediately. For Euler's proof that the formula is valid for any real n, see Smith, pp. 452–454.

4. Ibid., pp. 447–450. Two of the roots appearing there, $-3/2 + (5\sqrt{3})i/2$ and $-3 + (\sqrt{3})i/2$, are clearly wrong, probably as a result of a misprint.

5. The history of the cubic equation is a long one, replete with controversy and intrigue. See David Eugene Smith, *History of Mathematics* (1925; rpt. New York: Dover, 1958), vol. 2, pp. 454–466; Victor J. Katz, *A History of Mathematics: An Introduction* (New York: HarperCollins, 1993), pp. 328–337; and David M. Burton, *History of Mathematics: An Introduction* (Dubuque, Iowa: Wm. C. Brown, 1995), pp. 288–299.

6. Ronald W. Clark, *Einstein: The Life and Times* (1971; rpt. New York: Avon Books, 1972), p. 270.

6

Two Theorems from Geometry

It is the glory of geometry that from so few principles,

fetched from without, it is able to accomplish so much.

—Sir Isaac Newton, preface to the *Principia*

Proposition 20 of Book III of Euclid's *Elements* says:

In a circle the angle at the center is double of the angle at the circumference, when the angles have the same circumference as base.[1]

In more common language, the proposition says that an angle inscribed in a circle (that is, an angle whose vertex lies on the circumference) is equal to half the central angle that subtends the same chord (fig. 28). Two corollaries from this theorem immediately follow: (1) In a given circle, *all* inscribed angles subtending the same chord are equal (this is Proposition 21 of Euclid; see fig. 29); and (2) All inscribed angles subtending the diameter are right angles (fig. 30). This last result is said to have been proved by Thales (although the Babylonians had already known it a thousand years before him) and may be one of the earliest theorems ever to have been proved.

This simple theorem, with its two corollaries, is a treasure trove of trigonometric information, and we will have numerous occasions to use it throughout this book. Let us use it here to prove the Law of Sines. Figure 31 shows a triangle ABC inscribed in a circle with center at O and radius r. We have $\angle AOB = 2\angle ACB = 2\gamma$. Drop the perpendicular bisector from O to AB. Then $\sin \gamma = (c/2)/r$, hence $c/\sin \gamma = 2r = $ constant. Since the ratio $c/\sin \gamma$ is constant (i.e., has the same value regardless of c and γ), we have

$$\frac{a}{\sin \alpha} = \frac{\beta}{\sin \beta} = \frac{c}{\sin \gamma} = 2r. \tag{1}$$

Not only is this proof a model of simplicity, it also gives the sine law in its complete form; the more common proof, based on

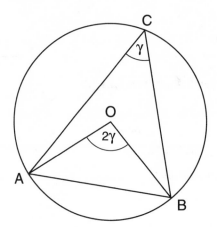

FIG. 28. Propositon 20 of Book III of Euclid's *Elements*.

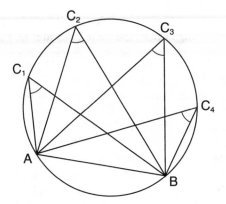

FIG. 29. Proposition 21 of Book III of Euclid's *Elements*.

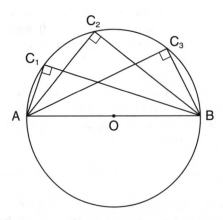

FIG. 30. All inscribed angles subtending a diameter are right angles.

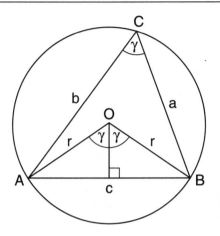

FIG. 31. The Law of Sines: the acute case.

dissecting a triangle into two right triangles, entirely ignores the statement about $2r$.

In figure 31 the angle γ is acute, which means that the center of the circle lies inside the triangle. If γ is obtuse (fig. 32), the center lies outside, so that arc AB is greater than half the circumference. Thus the internal angle of triangle AOB at O is $\gamma' = 360° - 2\gamma$. Again drop the perpendicular bisector from O to AB: we have $\sin \gamma'/2 = (c/2)/r$. But $\sin \gamma'/2 = \sin(180° - \gamma) = \sin \gamma$, so we again get $c/\sin \gamma = 2r$, as before.

We can get still more information out of our theorem. Figure 33 shows the unit circle and a point P on it. Let the angle between OP and the positive x-axis be 2θ. Then $\angle ORP = \theta$, where R is the point with coordinates $(-1, 0)$. Applying the sine law to the triangle ORP, we have $RP/\sin(180° - 2\theta) = OP/\sin \theta$. But $\sin(180° - 2\theta) = \sin 2\theta$ and $OP = 1$, so $RP/\sin 2\theta = 1/\sin \theta$,

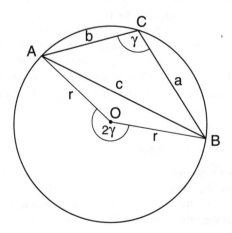

FIG. 32. The Law of Sines: the obtuse case.

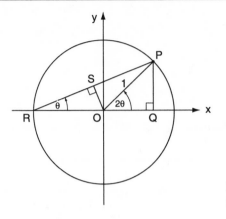

FIG. 33. Geometric proof of the double-angle formulas.

from which we get

$$\sin 2\theta = RP \sin \theta. \tag{2}$$

Now drop the perpendicular bisector OS from O to RP; in the right triangle ORS we have $\cos \theta = RS/RO = (RP/2)/RO = RP/2$, hence $RP = 2\cos \theta$. Substituting this back in equation (2), we get

$$\sin 2\theta = 2\sin \theta \cos \theta, \tag{3}$$

which is the double-angle formula for the sine. Again, dropping the perpendicular PQ from P to the x-axis, we have

$$\cos 2\theta = OQ = RQ - RO = RP \cos \theta - 1$$
$$= (2\cos \theta) \cdot \cos \theta - 1 = 2\cos^2 \theta - 1, \tag{4}$$

which is the double-angle formula for the cosine. Finally, having proved the double-angle formulas, we can derive the corresponding half-angle formulas by simply replacing 2θ by ϕ.

Let us return for a moment to our proof of the Law of Sines. Since any three noncollinear points determine a circle uniquely, every triangle can be inscribed in exactly one circle. Indeed, we may regard the angles of the triangle as inscribed angles and the sides as chords in that circle. Thus the Law of Sines is really a theorem about circles. If we let the *diameter* of the inscribing circle be 1 and call this circle the "unit circle," then the Law of Sines simply says that

$$a = \sin \alpha, b = \sin \beta, c = \sin \gamma,$$

that is, *each side of a triangle inscribed in a unit circle is equal to the sine of the opposite angle* (fig. 34). We could, in fact, *define* the sine of an angle as the length of the chord it subtends in the unit circle, and this definition would be as good as the traditional

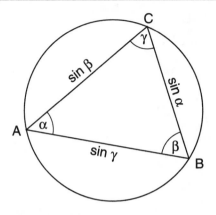

FIG. 34. The Law of Sines in a "unit circle."

definition of the sine as a ratio of two sides in a right triangle. (It would, in fact, have the advantage that the angle could vary from 0° to 180°—twice the range in a right triangle.) As we saw in chapter 2, it was this interpretation of the sine function that Ptolemy used in his table of chords.

✧   ✧   ✧

In Ptolemy's *Almagest* we find the following proposition, known as Ptolemy's Theorem:[2]

The rectangle contained by the diagonals of any quadrilateral inscribed in a circle is equal to the sum of the rectangles contained by the pairs of opposite sides.[3]

What is the meaning of this cryptic statement? To begin with, the Greeks interpreted a number as the length of a line segment, and the product of two numbers as the area of a rectangle with sides having the given numbers as lengths. Thus "the rectangle contained by the diagonals" means the area of a rectangle whose sides are the diagonals of an inscribed quadrilateral, with a similar interpretation for "the rectangles contained by the pairs of opposite sides." In short, "a rectangle contained by" simply means "a product of." Ptolemy's Theorem can then be formulated as follows: *In a quadrilateral inscribed in a circle, the product of the diagonals is equal to the sum of the products of the opposite sides.* Referring to figure 35, this means that

$$AC \cdot BD = AB \cdot CD + BC \cdot DA. \qquad (5)$$

As this theorem is not as widely known as others in elementary geometry, we give here Ptolemy's proof: Using one side, say AB, as the initial side, we construct an angle ABE equal to DBC. Now angles CAB and CDB are also equal, having the

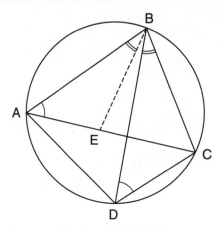

Fig. 35. Ptolemy's Theorem.

common chord BC. Therefore, triangles ABE and DBC are similar, having two pairs of equal angles. Hence $AE/AB = DC/DB$, from which we get

$$AE \cdot DB = AB \cdot DC. \tag{6}$$

If we now add the angle EBD to both sides of the equation $\angle ABE = \angle DBC$, we get $\angle ABD = \angle EBC$. But angles BDA and BCE are also equal, having the common chord AB. Therefore, triangles ABD and EBC are similar, hence $AD/DB = EC/CB$ and thus

$$EC \cdot DB = AD \cdot CB. \tag{7}$$

Finally, adding equations (6) and (7), we have $(AE + EC) \cdot DB = AB \cdot DC + AD \cdot CB$; replacing $AE + EC$ by AC, we get the required result (note that the sides are nondirected line segments, so that $BD = DB$, etc.).

If we let the quadrilateral $ABCD$ be a rectangle (fig. 36), then all four vertices form right angles, and furthermore $AB = CD$, $BC = DA$, and $AC = BD$. Equation (5) then says that

$$(AC)^2 = (AB)^2 + (BC)^2, \tag{8}$$

which is the Pythagorean Theorem! This demonstration of the most celebrated theorem of mathematics appears as number 66 of 256 proofs in Elisha Scott Loomis's classic book, *The Pythagorean Proposition*.[4]

What is the trigonometric significance of Ptolemy's Theorem? For the special case where $ABCD$ is a rectangle, AC is a diameter in our "unit circle," hence $AC = 1$. Moreover, denoting angle BAC by α, we have $AB = \cos \alpha$, $BC = \sin \alpha$. Equation (8)

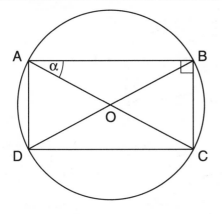

FIG. 36. The Pythagorean Theorem.

thus becomes

$$1 = \cos^2 \alpha + \sin^2 \alpha,$$

which is the trigonometric equivalent of the Pythagorean Theorem. But there is more in store. Let $ABCD$ be any quadrilateral in which one diagonal, say AC, coincides with the diameter (fig. 37). Then $\angle ABC$ and $\angle ADC$ are right angles. Let $\angle BAC = \alpha$, $\angle CAD = \beta$. We then have $BC = \sin \alpha$, $AB = \cos \alpha$, $CD = \sin \beta$, $AD = \cos \beta$, and $BD = \sin (\alpha + \beta)$, so that by Ptolemy's Theorem,

$$1 \cdot \sin (\alpha + \beta) = \sin \alpha \cdot \cos \beta + \cos \alpha \cdot \sin \beta,$$

which is the addition formula for the sine function! (The difference formula $\sin (\alpha - \beta) = \sin \alpha \cdot \cos \beta - \cos \alpha \cdot \sin \beta$ can likewise be obtained by considering a quadrilateral in which one *side*, say AD, coincides with the diameter; see fig. 38.) Thus what is perhaps the single most important formula in trigonometry was

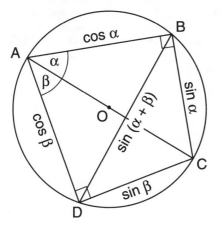

FIG. 37. Geometric proof of $\sin (\alpha + \beta) = \sin \alpha \cos \beta + \cos \alpha \sin \beta$.

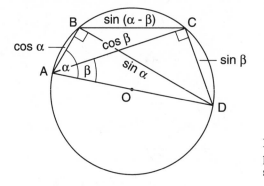

FIG. 38. Geometric proof of $\sin(\alpha - \beta) = \sin\alpha\cos\beta - \cos\alpha\sin\beta$.

already known to Ptolemy, who used it to great effect in calculating his table of chords; and quite possibly it was discovered already by Hipparchus two and a half centuries earlier. The old adage is still true: "Nothing is new under the sun."

NOTES AND SOURCES

1. Euclid, *The Elements*, translated with introduction and commentary by Sir Thomas Heath (Annapolis: St. John's College Press, 1947), vol. 2, pp. 46–49.

2. Tobias Dantzig, in his book *The Bequest of the Greeks* (New York: Charles Scribner's Sons, 1955), p. 173, suggests that the theorem might have been discovered by Apollonius, who lived three centuries before Ptolemy.

3. Euclid, *The Elements*, vol. 2, pp. 225–228.

4. Loomis, *The Pythagorean Propositon* (1940; rpt. Washington, D.C.: The National Council of Teachers of Mathematics, 1968), p. 66. None of the 256 proofs relies on trigonometry: "There are no trigonometric proofs [of the Pythagorean Theorem], because all the fundamental formulae of trigonometry are themselves based upon the truth of the Pythagorean Theorem. . . . Trigonometry *is* because the Pythagorean Theorem *is*" (p. 244). Among the proofs there is one (number 231) proposed by James A. Garfield in 1876, five years before he became president of the United States.

7

Epicycloids and Hypocycloids

The epicycle theory [of the motion of the planets], in the
definite form given by Ptolemy, stands out as the most
mature product of ancient astronomy.
—Anton Pannekoek, *A History of Astronomy*

In the 1970s an intriguing educational toy appeared on the market and quickly became a hit: the *spirograph*. It consisted of a set of small, plastic-made wheels of varying sizes with teeth along their rims, and two large rings with teeth on their inside as well as outside rims (fig. 39). Small holes perforated each wheel at various distances from the center. You pinned down one of the rings onto a sheet of paper, placed a wheel in contact with it so the teeth would engage, and inserted a pen in one of the holes. As you moved the wheel around the ring, a curve was traced on the paper—a *hypocycloid* if the wheel moved along the inside rim of the ring, an *epicycloid* if it moved along the outside rim (the names come from the Greek prefixes *hypo* = under, and *epi* = over). The exact shape of the curve depended on the radii of the ring and wheel (each expressed in terms of the number of teeth on its rim); more precisely, on the *ratio* of radii.

Let us find the parametric equations of the hypocycloid, the curve traced by a point on a circle of radius r as it rolls on the inside of a fixed circle of radius R (fig. 40). Let O and C be the centers of the fixed and rolling circles, respectively, and P a point on the moving circle. When the rolling circle turns through an angle ϕ in a clockwise direction, C traces an arc of angular width θ in a *counterclockwise* direction relative to O. Assuming that the motion starts when P is in contact with the fixed circle at the point Q, we choose a coordinate system in which the origin is at O and the x-axis points to Q. The coordinates of P relative to C are $(r\cos\phi, -r\sin\phi)$ (the minus sign in the second coordinate is there because ϕ is measured clockwise), while the coordinates of C relative to O are $((R-r)\cos\theta, (R-r)\sin\theta)$.

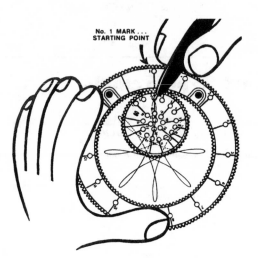

FIG. 39. Spirograph.

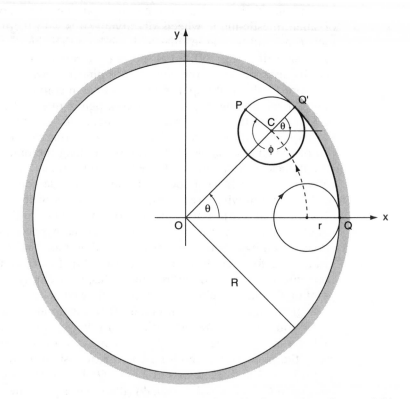

FIG. 40. Generating a hypocycloid.

Thus the coordinates of P *relative to* O are

$$x = (R - r)\cos\theta + r\cos\phi, \quad y = (R - r)\sin\theta - r\sin\phi. \quad (1)$$

But the angles θ and ϕ are not independent: as the motion progresses, the arcs of the fixed and moving circles that came in contact (arcs QQ' and $Q'P$ in fig. 40) must be of equal length. These arcs have lengths $R\theta$ and $r(\theta + \phi)$, respectively, so we have $R\theta = r(\theta + \phi)$. Using this relation to express ϕ in terms of θ, we get $\phi = [(R - r)/r]\theta$, so that equations (1) become

$$x = (R - r)\cos\theta + r\cos[(R - r)/r]\theta,$$
$$y = (R - r)\sin\theta - r\sin[(R - r)/r]\theta. \quad (2)$$

Equations (2) are the *parametric equations* of the hypocycloid, the angle θ being the parameter (if the rolling circle rotates with constant angular velocity, θ will be proportional to the elapsed time since the motion began). The general shape of the curve depends on the ratio R/r. If this ratio is a fraction m/n in lowest terms, the curve will have m cusps (corners), and it will be completely traced after moving the wheel n times around the inner rim. If R/r is irrational, the curve will never close, although going around the rim many times will nearly close it.

For some values of R/r the resulting curve may be something of a surprise. For example, when $R/r = 2$, equations (2) become

$$x = r\cos\theta + r\cos\theta = 2r\cos\theta,$$
$$y = r\sin\theta - r\sin\theta = 0. \quad (3)$$

The fact that $y = 0$ at all times means that P moves along the x-axis only, tracing the inner diameter of the ring back and forth. Thus we can use two circles of radii ratio 2:1 to draw a straight line segment! In the nineteenth century the problem of converting circular motion to rectilinear and vice versa was crucial to the design of steam engines: the to-and-fro movement of the piston had to be converted to a rotation of the wheels. The 2:1 hypocycloid was one of numerous solutions proposed.

Even more interesting is the case $R/r = 4$, for which equations (2) become

$$x = 3r\cos\theta + r\cos 3\theta,$$
$$y = 3r\sin\theta - r\sin 3\theta. \quad (4)$$

To get the rectangular equation of the curve—the equation connecting the x- and y-coordinates of P—we must eliminate the parameter θ between the two equations. Generally this may require some tedious algebraic manipulations, and the resulting equation—if it can be obtained at all—can be very complicated.

But in this case a pair of trigonometric identities come to our help—the identities $\cos^3 \theta = (3\cos\theta + \cos 3\theta)/4$ and $\sin^3 \theta = (3\sin\theta - \sin 3\theta)/4$.[1] Equations (4) then become

$$x = 4r\cos^3 \theta, \; y = 4r\sin^3 \theta.$$

Taking the cube root of each equation, squaring the results, and adding, we finally get

$$x^{2/3} + y^{2/3} = (4r)^{2/3} = R^{2/3}. \tag{5}$$

The hypocycloid described by equation (5) is called an *astroid*; it has the shape of a star (hence the name) with four cusps located at $\theta = 0°, 90°, 180°,$ and $270°$. The astroid has some remarkable properties. For example, all its tangent lines intercept the same length between the axes, this length being R. And conversely, if a line segment of fixed length R and endpoints on the x- and y-axes is allowed to assume all possible positions, the *envelope* formed by all the line segments—the curve tangent to every one of them—is an astroid (fig. 41). Hence the region occupied by a ladder leaning against a wall as it assumes all possible positions has the shape of an astroid. Surprisingly, the astroid is also the envelope of the family of ellipses $x^2/a^2 + y^2/(R-a)^2 = 1$, the sum of whose semimajor and semiminor axes is R (fig. 42).[2]

It so happens that the rectangular equation of the astroid (equation 5) makes it particularly easy to compute various metric properties of this curve. For example, using a formula from calculus for finding the arc length of a curve, one can show that the circumference of the astroid is $6R$ (surprisingly, despite the

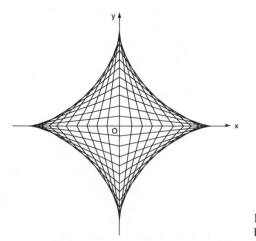

FIG. 41. Astroid formed by its tangent lines.

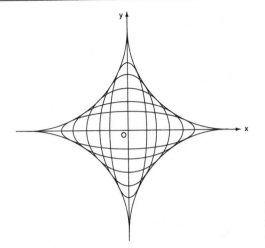

FIG. 42. Astroid formed by tangent ellipses.

involvement of circles in generating the astroid, its circumference does not depend on the constant π). The area enclosed by the astroid is $3\pi R^2/8$, or three-eighths the area of the fixed circle.[3]

✧ ✧ ✧

In 1725 Daniel Bernoulli (1700–1782), a member of the venerable Bernoulli family of mathematicians, discovered a beautiful property of the hypocycloid known as the *double generation theorem*: a circle of radius r rolling on the inside of a fixed circle of radius R generates the same hypocycloid as does a circle of radius $(R - r)$ rolling inside the same fixed circle. If we denote the former hypocycloid by $[R, r]$ and the latter by $[R, R - r]$, the theorem says that $[R, r] = [R, R - r]$. Note that the two rolling circles are complements of each other with respect to the fixed circle: the sum of their diameters equals the diameter of the fixed circle (fig. 43).

To prove this theorem, let us take advantage of a peculiar skew-symmetry in equations (1). Substituting $r' = R - r$ in these equations, we get

$$x = r' \cos \theta + (R - r') \cos \phi, \, y = r' \sin \theta - (R - r') \sin \phi.$$

But the parameters θ and ϕ are related through the equation $(R - r)\theta = r\phi$. Using this equation to express θ in terms of ϕ, we have $\theta = r\phi/(R - r) = [(R - r')/r']\phi$. Equations (1) thus become

$$x = r' \cos [(R - r')/r']\phi + (R - r') \cos \phi,$$
$$y = r' \sin [(R - r')/r']\phi - (R - r') \sin \phi. \tag{6}$$

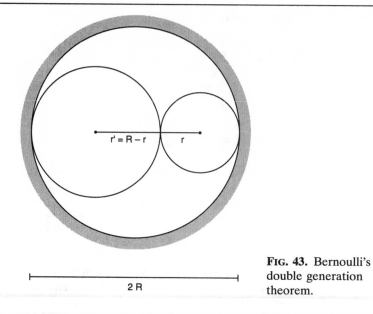

$r' = R - r$ r

2 R

FIG. 43. Bernoulli's double generation theorem.

Equations (6), except for the fact that r' replaces r, are strikingly similar to equations (2). Indeed, we can make them *identical* with equations (2) by interchanging the order of terms in each equation:

$$x = (R - r')\cos\phi + r'\cos[(R - r')/r']\phi,$$
$$y = -(R - r')\sin\phi + r'\sin[(R - r')/r']\phi.$$

The first of these equations is exactly identical with the first of equations (2), with r' replacing r and θ and ϕ interchanged.[4] But the second equation still has a bothersome misplacement of signs: we would like the first term to be positive and the second term negative. Here again a pair of trigonometric identities come to our help, the even-odd identities $\cos(-\phi) = \cos\phi$ and $\sin(-\phi) = -\sin\phi$. So let us change our parameter once more, replacing ϕ by $\psi = -\phi$; this leaves the terms of the first equation unaffected but interchanges the signs of the terms in the second equation:

$$x = (R - r')\cos\psi + r'\cos[(R - r')/r']\psi,$$
$$y = (R - r')\sin\psi - r'\sin[(R - r')/r']\psi,$$

(7)

which are identical with equations (2). This completes the proof.[5]

As a consequence of this theorem we have, for example, $[4r, r] = [4r, 3r]$—or, equivalently, $[R, R/4] = [R, 3R/4]$—showing that the astroid described by equation (5) can also be generated by a circle of radius $3R/4$ rolling on the inside of a fixed circle of radius R.

✧ ✧ ✧

The parametric equations of the epicycloid—the curve traced by a point on a circle of radius r rolling on the outside of a fixed circle of radius R—are analogous to those of the hypocycloid (equations (2)):

$$x = (R+r)\cos\theta - r\cos[(R+r)/r]\theta,$$
$$y = (R+r)\sin\theta - r\sin[(R+r)/r]\theta. \tag{8}$$

The appearance of $(R+r)$ instead of $(R-r)$ is self-explanatory, but note also the minus sign in the second term of the x-equation; this is because the rotation of the rolling circle and the motion of its center are now in the *same* direction.

As with the hypocycloid, the shape of the epicycloid depends on the ratio R/r. For $R/r = 1$ equations (8) become $x = r(2\cos\theta - \cos 2\theta)$, $y = r(2\sin\theta - \sin 2\theta)$, and the resulting curve is the heart-shaped cardioid (fig. 44). It has a single cusp, located where P comes in contact with the fixed circle. Its circumference is $16R$ and its area $6\pi R^2$.[6]

One more case must be considered: a circle of radius r rolling on the outside of a fixed circle of radius R touching it *internally* (fig. 45).[7] This case is similar to the hypocycloid, except that the roles of the fixed and rolling circles are reversed. The parametric equations in this case are

$$x = r\cos\phi - (r-R)\cos\theta, \, y = r\sin\phi - (r-R)\sin\theta,$$

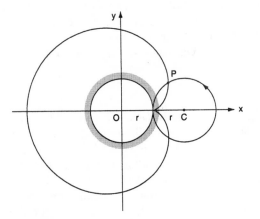

FIG. 44. Cardioid.

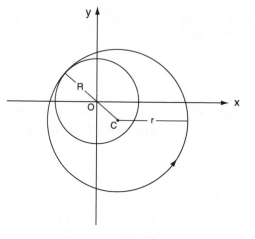

FIG. 45. A large circle rolling on the outside of a small circle, touching it internally.

(note that now $r > R$) where θ and ϕ are related through the equation $(r - R)\theta = r\phi$. Expressing θ in terms of ϕ and making the substitution $r' = r - R$, these equations become

$$x = (R + r') \cos \phi - r' \cos \left[(R + r')/r'\right]\phi$$
$$y = (R + r') \sin \phi - r' \sin \left[(R + r')/r'\right]\phi. \tag{9}$$

Equations (9) are identical with equations (8) for the epicycloid, except that r is replaced by r' and θ by ϕ. The ensuing curve is therefore identical with an epicycloid generated by a circle of radius $r' = r - R$ rolling on the outside of a fixed circle of radius R touching it *externally*. And conversely, the latter epicycloid is identical with the curve generated by a circle of radius $r = R + r'$ rolling on the outside of a fixed circle of radius R touching it internally. This is the double generation theorem for epicycloids. If we introduce the symbols { } and ( ) to denote the "external" and "internal" epicycloids, respecivtely, the theorem says that $\{R, r\} = (R, R + r)$ (we have dropped the prime over the r). Thus for the cardioid we have $\{R, R\} = (R, 2R)$.

✧   ✧   ✧

The study of epicycloids goes back to the Greeks, who used them to explain a puzzling celestial phenomenon: the occasional *retrograde motion* of the planets as viewed from the earth. During most of its motion along the zodiac a planet moves from west to east; but occasionally the planet seems to come to a standstill, reverse its course to an east-to-west motion, then stop again and resume its normal course. To the aesthetically minded Greeks, the only imaginable curve along which the heavenly bodies could

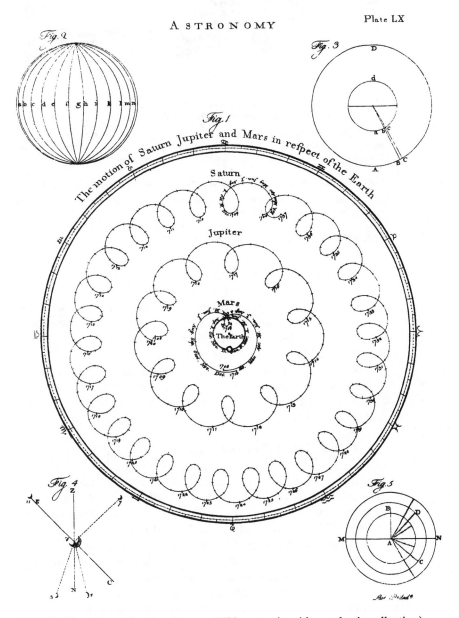

FIG. 46. Planetary epicycles. From a 1798 engraving (the author's collection).

No. 1181.

1179. Ellipsograph, brass, nickelplated, fine quality, 6 in. bar,
   with pen and pencil point (in one piece). In case, . . . each $

This instrument draws ellipses of any shape, from 4 inches to 11 inches
major axis, with great accuracy. Its construction is shown by the illustration.
The pen-pencil point can be taken off and stored compactly in the case.

1181. Ellipsograph, like No. 1179, but with 9 in. bar. In case, . . each $

This instrument draws ellipses of any shape, from 6 inches to 18 inches
major axis, with great accuracy.

FIG. 47. Ellipsograph. From Keuffel & Esser's catalog, 1928.

move around the earth was the circle—the symbol of perfection.
But a circle does not allow for retrograde motion, so the Greeks
assumed that the planet actually moves along a small circle, the
epicycle, whose center moves along the main circle, the *defer-
ent* (fig. 46). When even this model did not quite account for
the observed motion of the planets, they added more and more
epicycles, until the system was so encumbered with epicycles as
to become unwieldy. Nevertheless, the system did describe the
observational facts at least approximately and was the first truly
mathematical attempt to account for the motion of the celestial
bodies.

It was only when Copernicus published his heliocentric theory
in 1543 that the need for epicycles disappeared: with the earth
orbiting the sun, the retrograde motion was at once explained
as a consequence of the relative motion of the planet as seen
from the moving earth. So when the Danish astronomer Olaus
Roemer (1644–1710), famous as the first to determine the speed
of light, undertook to investigate cycloidal curves in 1674, it was
in connection not to the heavenly bodies but to a more mundane
problem—the working of mechanical gears.

With today's computers and graphing calculators, one can
generate even the most complex curves within seconds. But
only a generation or two ago, such a task relied entirely on mechanical devices; indeed, a number of ingenious instruments
were invented to draw specific types of curves (figs. 47 and 48).[8]
Often these devices involved highly complex mechanisms, but

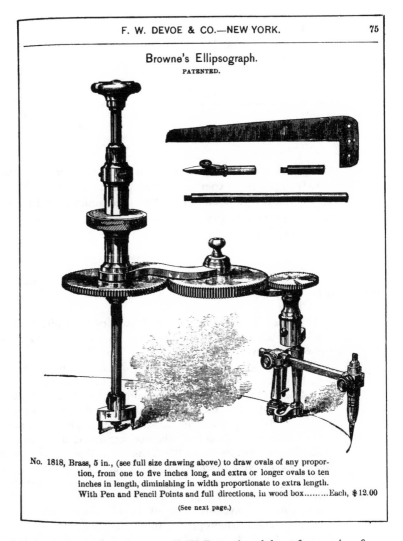

FIG. 48. Ellipsograph. From F. W. Devoe's calalog of surveying & mathematical instruments, ca. 1900.

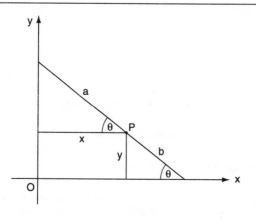

FIG. 49. As θ varies, P describes an arc of an ellipse.

there was a certain fascination in watching the gears move and slowly trace the expected curve; you could, quite literally, see the machine at work. With the mechanical world giving way to the electronic era, efficiency triumphed at the expense of intimacy.[9]

NOTES AND SOURCES

1. These identities are obtained by solving the triple-angle formulas $\cos 3\theta = 4\cos^3\theta - 3\cos\theta$ and $\sin 3\theta = 3\sin\theta - 4\sin^3\theta$ for $\cos^3\theta$ and $\sin^3\theta$, respectively.

2. To see this, consider a fixed point $P(x, y)$ on a line segment of length R whose endpoints are free to move along the x- and y-axes (fig. 49). If P divides the segment into two parts of lengths a and b, we have $\cos\theta = x/a$, $\sin\theta = y/b$. Squaring and adding, we get $x^2/a^2 + y^2/b^2 = 1$, the equation of an ellipse with semimajor axis a and semiminor axis $b = R - a$. Thus when the line segment is allowed to assume all possible positions, the point P will trace the ellipse (this is the basis for the ellipse-drawing machine shown in fig. 47). For different positions of P along the line segment (i.e., when the ratio a/b assumes different values, while $a + b$ is kept constant) different ellipses will be drawn, whose common envelope is the astroid $x^{2/3} + y^{2/3} = R^{2/3}$.

3. For additional properties of the astroid, see Robert C. Yates, *Curves and their Properties* (1952; rpt. Reston, Virginia: National Council of Teachers of Mathematics, 1974), pp. 1–3.

4. Note that we are free to replace one parameter by another, provided that the new parameter causes x and y to cover the same range of values as the old. In our case this is ensured by the periodicity of the sine and cosine functions.

5. The double generation theorm can also be proved geometrically; see, Yates, *Curves*, pp. 81–82.

6. The familiar polar equation of the cardioid, $\rho = r(1 - \cos\theta)$, holds when the cusp is at the origin (here θ denotes the polar angle between the positive x-axis and the line OP; this is not to be confused with the angle θ appearing in equations 8). For additional properties of the cardioid, see Yates, *Curves*, pp. 4–7.

7. I am indebted to Robert Langer of the University of Wisconsin–Eau Claire for calling my attention to this case.

8. See H. Martyn Cundy and A. P. Rollett, *Mathematical Models* (London: Oxford University Press, 1961), chapters 2 and 5.

9. The visitor to the Museum of Science and Industry in Chicago will find an interesting display of mechanical gears, modestly tucked along one of the stairways and almost eclipsed by the larger exhibits that fill the museum's cavernous halls. Moving a small crank with your hand, you can activate the gears and watch the ensuing motion—a mute reminder of a bygone era.

Maria Agnesi and Her "Witch"

Even today, women make up only about 10 percent of the total number of mathematicians in the United States;[1] worldwide their number is much smaller. But in past generations, social prejudices made it almost impossible for a woman to pursue a scientific career, and the total number of women mathematicians up to our century can be counted on two hands. Three names come to mind: Sonia Kovalevsky (1850–1891) of Russia, Emmy Noether (1882–1935), who was born in Germany but emigrated to the U.S., and Maria Agnesi of Italy.[2]

Maria Gaetana Agnesi (pronounced "Anyesi") was born in Milan in 1718, where she spent most of her life.[3] Her father, Pietro, a wealthy professor of mathematics at the University of Bologna, encouraged her to study the sciences. To further her education he founded at their home a kind of "cultural salon" where guests would come from all over Europe, many of them scholars in various fields. Before these guests, young Maria displayed her intellectual talents by presenting theses on a variety of subjects and then defending them in disputation. The subjects included logic, philosophy, mechanics, chemistry, botany, zoology and mineralogy. During intermission her sister, Maria Teresa, who was a composer and harpsichordist, entertained the guests with her music. The scene is reminiscent of Leopold Mozart showing off young Amadeus's musical talents in the salons of the well-to-do of Salzburg, with Mozart's sister Nannerl playing in the background. Maria Gaetana was also versed in languages: at the age of five she was already fluent in French, and when she was nine she translated into Latin and published a long speech advocating higher education for women. Soon she mastered Greek, German, Spanish, and Hebrew and would defend her theses in her guests' native languages. She later collected 190 of these theses and published them in a book, *Proportiones philosophicae* (1738); unfortunately, none of her mathematical thoughts are included in this work.

By the time Agnesi was fourteen she was already solving difficult problems in analytic geometry and physics. At seventeen she began shaping her critical commentary on Guillaume L'Hospital's work, *Traité analytique des sections coniques*; unfortunately, this commentary was never published. About that time she had had enough with the public displays of her talents; she withdrew from social life to devote herself entirely to mathematics. She spent the next ten years writing her major work, *Instituzioni analytiche ad uso della gioventu italiana* (Analytic institutions for the use of young Italians). This work was published in 1748 in two very large volumes, the first dealing with algebra and the second with analysis (that is, infinite processes). Her goal was to give a complete and integrated presentation of these subjects as they were then known (we must remember that in the middle of the eighteenth century the calculus was still in a developmental stage, and new procedures and theorems were constantly being added to its existing core). Agnesi wrote her book in Italian rather than Latin, the scholarly language of the time, in order to make it accessible to as many "young Italians" as possible.

The *Instituzioni* brought Agnesi immediate recognition and was translated into several languages. John Colson (d. 1760), Lucasian professor at Cambridge University, who in 1736 published the first full exposition of Newton's *Method of Fluxions and Infinite Series* (his differential calculus), translated Agnesi's book into English. This he did when he was already advanced in age, learning Italian expressly for this task "so that the British Youth might have the benefit of it as well as the Youth of Italy." His translation was published in London in 1801.

In recognition of her accomplishments, Pope Benedict XIV in 1750 appointed Agnesi professor of mathematics at the University of Bologna. But she never actually taught there, viewing her position merely as an honorary one. After her father's death in 1752 she gradually withdrew from scientific activity, devoting her remaining years to religious and social work. She also raised her father's twenty-one children (from three marriages) and directed their education, while at the same time helping the poor in her parish. She died in Milan in 1799 at the age of 81.

✧ ✧ ✧

It is ironic that Agnesi's name is mainly remembered today for a curve she has investigated but was not the first to study: the *witch of Agnesi*. Consider a circle of radius a and center at $(0, a)$ (fig. 50). A line through $(0, 0)$ cuts the circle at point A and is extended until it meets the horizontal line $y = 2a$ at point B. Draw a horizontal line through A and a vertical line through B, and let these lines meet at P. The witch is the locus of P as the line OA assumes all possible positions.

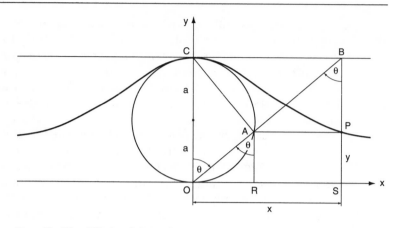

FIG. 50. The Witch of Agnesi.

It is easiest to find the equation of the witch in terms of the angle θ between OA and the y-axis. Let the coordinates of P be (x, y). From figure 50 we see that $\angle OAC = 90°$, OC being a diameter of the circle; in the right triangle OAC we thus have $OA = OC \cos \theta = 2a \cos \theta$. Let R and S be the feet of the perpendiculars from A and B to the x-axis, respectively; then in the right triangle OBS we have $OS = x = BS \tan \theta = 2a \tan \theta$, and in the right triangle OAR we have $AR = y = OA \cos \theta = 2a \cos^2 \theta$. The parametric equations of the witch are thus

$$x = 2a \tan \theta, \quad y = 2a \cos^2 \theta. \tag{1}$$

To find its rectangular equation we must eliminate θ between equations (1). Using the identity $1 + \tan^2 \theta = 1/\cos^2 \theta$ to express y in terms of x, we get

$$y = \frac{8a^3}{x^2 + 4a^2}. \tag{2}$$

Several conclusions follow from equation (2). First, as $x \to \pm\infty$, $y \to 0$, showing that the x-axis is a horizontal asymptote of the witch. Secondly, using calculus one can show that the area between the witch and its asymptote is $4\pi a^2$, or four times the area of the generating circle.[4] It can also be shown—either directly from equation (2) or from the parametric equations (1)—that the witch has two inflection points (points where the curve changes its concavity), located at $\theta = \pm\pi/6$. The calculation is a bit lengthy but otherwise straightforward, and we will omit it.[5]

As already mentioned, the "witch" did not originate with Agnesi; it was already known to Pierre Fermat (1601–1665), and Luigi Guido Grandi (1671–1742), a professor of mathematics at the University of Pisa, gave it the name *versiera* (from the

Latin *vertere*, to turn). It so happened, however, that a similar-sounding Italian word, *avversiera* means a female devil or devil's wife. According to D. J. Struik, "Some wit in England once translated it 'witch,' and the silly pun is still lovingly preserved in most of our textbooks in the English language."[6] So Grandi's *versiera* became "Agnesi's witch." It is somewhat of a mystery why this particular curve, which rarely shows up in applications, has interested mathematicians for so long.[7] Its strange name may have something to do with it, or perhaps it was Agnesi's role in making the curve known.

Notes and Sources

1. This figure is based on the Annual *AMS-IMS-MAA* Survey, Notices of the American Mathematical Society, Fall 1993.

2. A good source on women scientists is Marilyn Bailey Ogilvie, *Women in Science—Antiquity through the Nineteenth Century: A Biographical Dictionary with Annotated Bibliography* Cambridge, Mass.: MIT Press, 1988). See also Lynn M. Osen, *Women in Mathematics* (1974; rpt. Cambridge, Mass.: MIT Press, 1988), and *Women of Mathematics: A Bibliographical Sourcebook*, ed. Louise S. Grinstein and Paul J. Campbell (New York: Greenwood Press, 1987).

3. The biographical details in this chapter are adapted from the *DSB*, vol. 1, pp. 75–77. See also Ogilvie, *Women in Science*, pp. 26–28.

4. This follows from the formula

$$2 \int_0^\infty y\, dx = 8a^2 \tan^{-1}(x/2a) \Big|_0^\infty = 4\pi a^2,$$

where $\tan^{-1}$ is the inverse tangent (or arctangent) function; we have used the fact that the witch is symmetric about the y-axis.

5. Some additional properties of the witch can be found in Robert C. Yates, *Curves and their Properties* (Reston, Va.: National Council of Teachers of Mathematics, 1974), pp. 237–238.

6. *A Source Book in Mathematics: 1200–1800* (Cambridge, Mass.: Harvard University Press, 1969), pp. 178–180. According to this source, the first to use the name "witch" in this sense may have been B. Williamson in his *Integral Calculus* (1875). Yates (in *Curves*, p. 237) has a different version of the evolution of the name "witch": "It seems Agnesi confused the old Italian word 'versorio' (the name given to the curve by Grandi), which means 'free to move in any direction', with 'versiera', which means 'goblin', 'bugaboo', 'Devil's wife', etc."

7. The curve does show up in probability theory as the Cauchy distribution $f(x) = 1/\pi(1 + x^2)$, whose equation, apart from the constants, is identical with that of the witch.

8

Variations on a Theme by Gauss

*The solving of an astronomical problem [proposed by
the French Academy of Sciences in 1735], for which
several eminent mathematicians had demanded several
months' time. . . was solved by the illustrious Gauss in
one hour.*
—Florian Cajori, quoted in R. E. Moritz, *On
Mathematics and Mathematicians, p. 155*

There is a story about the great German mathematician Carl
Friedrich Gauss (1777–1855), who as a schoolboy was asked by
his teacher to sum up the numbers from 1 to 100, and who
almost immediately came up with the correct answer, 5,050. To
the amazed teacher Gauss explained that he had merely noticed
that by writing the numbers twice, first from 1 to 100 and then
from 100 to 1, and adding the two sums vertically, each pair
adds up to 101. Since there are one hundred such pairs, we get
$100 \times 101 = 10,100$, and since this is twice the required sum, the
answer is one half of this, namely 5,050.

Like so many stories about famous people, this one may or
may not have actually happened; however, what matters is the
lesson to be drawn from it—the importance of looking for patterns. The pattern in this case is that of a staircase, where what
we add at one end is subtracted at the other end:

$$S = 1 + 2 + 3 + \cdots + n$$
$$S = n + (n-1) + (n-2) + \cdots + 1$$
$$2S = (n+1) + (n+1) + (n+1) + \cdots + (n+1) = n(n+1)$$

$$n \text{ terms}$$

$$S = n(n+1)/2. \tag{1}$$

I remembered the story about Gauss while browsing one day
through a handbook of sequences and series, where I found the

following summation formula:[1]

$$\sin \alpha + \sin 2\alpha + \sin 3\alpha + \cdots + \sin n\alpha$$
$$= \frac{\sin n\alpha/2 \cdot \sin (n+1)\alpha/2}{\sin \alpha/2}. \tag{2}$$

Not having at first any clue as to how to prove this formula, I began to look for a pattern. What struck me was the formal similarity between equations (1) and (2); indeed, multiplying both sides of equation (1) by α gives us $S\alpha = n(n+1)\alpha/2$; that is,

$$\alpha + 2\alpha + 3\alpha + \cdots + n\alpha = n(n+1)\alpha/2.$$

"Multiplying" this last equation by "sin" and proceeding as if "sin" were an ordinary algebraic quantity, we get

$$\sin (\alpha + 2\alpha + 3\alpha + \cdots + n\alpha) = \sin n(n+1)\alpha/2.$$

If on the left side we open the parentheses, and on the right side multiply and divide by a second "sin" (squeezing it between the n and $(n+1)$) and again by $\alpha/2$, we get equation (2)!

Of course, we have committed every imaginable mathematical *sin* (no pun intended), and yet we did get a correct formula. Can we, then, prove equation (2) in a manner similar to Gauss's handling of the sum in equation (1)?

Let

$$S = \sin \alpha + \sin 2\alpha + \cdots + \sin (n-1)\alpha + \sin n\alpha$$
$$S = \sin n\alpha + \sin (n-1)\alpha + \cdots + \sin 2\alpha + \sin \alpha.$$

Summing the terms vertically in pairs and using the sum-to-product formula $\sin \alpha + \sin \beta = 2 \sin (\alpha + \beta)/2 \cdot \cos (\alpha - \beta)/2$, we get

$$2S = 2[\sin (1+n)\alpha/2 \cdot \cos (1-n)\alpha/2 + \sin (1+n)\alpha/2$$
$$\cdot \cos (3-n)\alpha/2 + \cdots + \sin (n+1)\alpha/2$$
$$\cdot \cos (n-3)\alpha/2 + \sin (n+1)\alpha/2 \cdot \cos (n-1)\alpha/2]$$
$$= 2\sin (n+1)\alpha/2 \cdot [\cos (1-n)\alpha/2 + \cos (3-n)\alpha/2 + \cdots$$
$$+ \cos (n-3)\alpha/2 + \cos (n-1)\alpha/2].$$

To get rid of the bothersome $1/2, 3/2, \ldots$ appearing in the cosine terms, let us multiply the last equation by $\sin \alpha/2$ and use the product-to-sum formula $\sin \alpha \cdot \cos \beta = (1/2)[\sin (\alpha - \beta) +$

$\sin(\alpha + \beta)$]; we get

$$2S \sin \alpha/2 = \sin(n+1)\alpha/2$$

$$\cdot [\sin n\alpha/2 + \sin(1-n/2)\alpha + \sin(-1+n/2)\alpha$$

$$+ \sin(2-n/2)\alpha + \cdots + \sin(2-n/2)\alpha$$

$$+ \sin(-1+n/2)\alpha + \sin(1-n/2)\alpha$$

$$+ \sin n\alpha/2].$$

But $\sin(-1+n/2)\alpha = -\sin(1-n/2)\alpha$, and similarly for the other terms; the expression in the brackets is therefore a "telescopic sum," all of whose terms except the first and last (which are equal) cancel out. We thus have

$$2S \sin \alpha/2 = 2 \sin(n+1)\alpha/2 \cdot \sin n\alpha/2$$

or

$$S = \frac{\sin(n+1)\alpha/2 \cdot \sin n\alpha/2}{\sin \alpha/2},$$

which is the formula we wished to prove.

An analogous formula for the summation of cosines can be proved in a similar way:[2]

$$\cos \alpha + \cos 2\alpha + \cdots + \cos n\alpha$$

$$= \frac{\cos(n+1)\alpha/2 \cdot \sin n\alpha/2}{\sin \alpha/2}. \tag{3}$$

If we now divide equation (2) by equation (3), we get the beautiful formula

$$\tan(n+1)\alpha/2 = \frac{\sin \alpha + \sin 2\alpha + \cdots + \sin n\alpha}{\cos \alpha + \cos 2\alpha + \cdots + \cos n\alpha}. \tag{4}$$

But this is not the end of the story. Following the idea that every trigonometric formula ultimately derives from geometry, we turn to fig. 51. Starting at the origin (which for our purpose we have labeled as P_0), we draw a line segment $P_0 P_1$ of unit length forming an angle α with the positive x-axis. At P_1 we draw a second line segment of unit length forming an angle α with the first segment and therefore the angle 2α with the x-axis. Continuing in this manner n times, we arrive at the point P_n, whose coordinates we shall denote by X and Y. Clearly X is the sum of the horizontal projections of the n line segments, and Y is the sum of their vertical projections, so we have

$$X = \cos \alpha + \cos 2\alpha + \cdots \cos n\alpha,$$
$$Y = \sin \alpha + \sin 2\alpha + \cdots + \sin n\alpha. \tag{5}$$

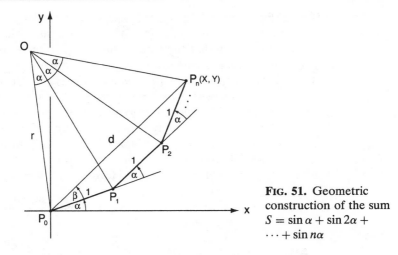

FIG. 51. Geometric construction of the sum $S = \sin \alpha + \sin 2\alpha + \cdots + \sin n\alpha$

Now the points P_i lie on a regular polygon inscribed in a circle with center O and radius r. Each line segment $P_{i-1}P_i$ subtends an angle α at O, and thus the segment P_0P_n subtends the angle $(n\alpha)$ at O. But this segment is the diagonal connecting the points P_0 and P_n; let us denote its length by d. In the isosceles triangle P_0OP_n we have

$$d = 2r \sin n\alpha/2,$$

while in the isosceles triangle P_0OP_1 we have

$$1 = 2r \sin \alpha/2.$$

Eliminating r between these two equations, we get

$$d = \frac{\sin n\alpha/2}{\sin \alpha/2}.$$

In order to find the horizontal and vertical projections of the segment P_0P_n, we need to find the angle it forms with the x-axis. This angle is $(\alpha + \beta)$, where $\beta = \angle P_1P_0P_n$. Now angle β subtends the chord P_1P_n in the inscribing circle and is therefore equal to half the central angle that subtends the same chord, that is, to $(n-1)\alpha/2$. Thus $\alpha + \beta = \alpha + (n-1)\alpha/2 = (n+1)\alpha/2$. Hence

$$X = d \cos (n+1)\alpha/2 = \frac{\cos (n+1)\alpha/2 \cdot \sin n\alpha/2}{\sin \alpha/2}$$

and   (6)

$$Y = d \sin (n+1)\alpha/2 = \frac{\sin (n+1)\alpha/2 \cdot \sin n\alpha/2}{\sin \alpha/2}.$$

If we substitute the expressions for X and Y from equations (5) into equations (6), we get equations (2) and (3).

If we think of each line segment $P_{i-1}P_i$ as a vector from P_{i-1} to P_i, then the line segment P_0P_n is their vector sum. Equations (2) and (3) then say that the sum of the (vertical or horizontal) projections of the individual line segments equals the (vertical or horizontal) projection of their vector sum. This shows that projection is a *linear operation*—an operation that obeys the distributive law $p(u + v) = p(u) + p(v)$, where $p(\ )$ stands for "projection of" and u and v are any two vectors. Projection—like all linear operations—behaves exactly as ordinary multiplication.

We can use "Gauss's method" of summation to prove other trigonometric summation formulas. Here are a few examples:

$$\sin \alpha + \sin 3\alpha + \sin 5\alpha + \cdots + \sin (2n - 1)\alpha = \frac{\sin^2 n\alpha}{\sin \alpha} \qquad (7)$$

$$\cos \alpha + \cos 3\alpha + \cos 5\alpha + \cdots + \cos (2n - 1)\alpha = \frac{\sin 2n\alpha}{2 \sin \alpha} \qquad (8)$$

$$\sin \pi/n + \sin 2\pi/n + \cdots + \sin n\pi/n = \cot \pi/2n \qquad (9)$$

$$\cos \pi/n + \cos 2\pi/n + \cdots + \cos n\pi/n = -1 \qquad (10)$$

$$\cos \pi/(2n + 1) + \cos 3\pi/(2n + 1) + \cdots$$
$$+ \cos (2n - 1)\pi/(2n + 1) = \tfrac{1}{2}. \qquad (11)$$

The last two are special cases of equations (3) and (8), respectively; they are remarkable because the sum in each is independent of n.

Trigonometric sums were studied by the Hungarian mathematician Lipót Fejér (1880–1959) in connection with his work on the summation of Fourier series, a subject to which we will return in chapter 15.

Notes and Sources

1. *Summation of Series*, collected by L.B.W. Jolley (1925; rpt. New York: Dover, 1961), series no. 417.

2. Both formulas can also be proved by taking the real and imaginary parts of the sum of the geometric progression $e^{i\alpha} + e^{2i\alpha} + \cdots + e^{ni\alpha}$, where $i = \sqrt{-1}$; see Richard Courant, *Differential and Integral Calculus* (1934; rpt. London: Blackie & Son, 1956), vol. 1, p. 436.

Had Zeno Only Known This!

One, Two, Three—Infinity
—Title of a book by George Gamow

Can space be endlessly divided, or is there a smallest unit of space, a mathematical atom that cannot further be split? Is motion continuous, or is it but a succession of snapshots which, like the frames in an old motion picture, are themselves frozen in time? Questions such as these were hotly debated by the philosophers of ancient Greece, and they are still being debated today—witness the never-ending search for the ultimate elementary particle, that elusive building block from which all matter is supposedly made.

The Greek philosopher Zeno of Elea, who lived in the fifth century B.C., summarized these questions in four paradoxes—he called them "arguments"—whose purpose was to demonstrate the fundamental difficulties inherent in the notion of continuity. In one of these paradoxes, known as the "dichotomy," he purports to show that motion is impossible: in order for a runner to go from point A to point B, he must first cover half the distance between A and B, then half the remaining distance, then half of that, and so on, *ad infinitum* (fig. 52). Since this involves an infinite number of steps, Zeno argued, the runner will never reach his destination.[1]

It is easy to formulate Zeno's paradox in modern terms. Let the distance from A to B be 1; by first covering half this distance, then half of what remains, and so on, the runner will cover a total distance given by the sum

$$1/2 + 1/4 + 1/8 + 1/16 + \cdots.$$

This sum is an endless geometric progression, or series, with the common ratio 1/2. As we add more and more terms, the sum keeps growing and approaches 1. It will never *reach* 1, let alone exceed 1; yet we can make the sum get as close to 1 as we please by simply adding more and more terms. In modern

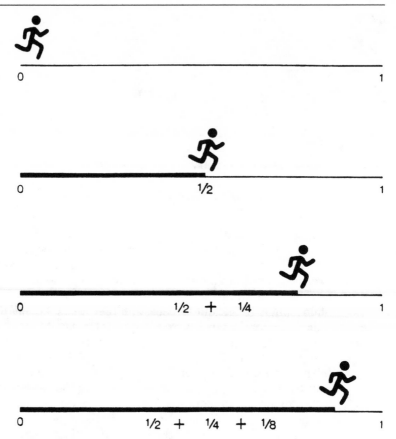

FIG. 52. The runner's paradox.

language, the sum approaches the *limit* 1 as the number of terms grows beyond bound. Thus the total distance covered is exactly 1; and since the time intervals it takes the runner to cover the partial distances (assuming he maintains a constant speed) also follow the same progression, he will cover the entire distance in finite time. This settles the "paradox."

The Greeks, however, did not subscribe to this kind of reasoning. They could not accept the fact—so obvious to us today—that a sum of infinitely many numbers may have a finite value. They had no difficulty adding up as many terms of a progression as were necessary to achieve a desired accuracy, but the thought of extending this process to infinity caused them great intellectual anguish. This in turn led to their *horror infiniti*—their fear of the infinite. Unable to deal with it, the Greeks barred infinity from their mathematical system. Although they had a firm in-

tuitive grasp of the limit concept—as evidenced by Archimedes'
quadrature of the parabola—they recoiled at the thought of going all the way to infinity.[2] As a result, Zeno's paradoxes remained a source of irritation and embarrassment to generations
of scholars. Frustrated by their failure to resolve the paradoxes
satisfactorily, they turned to philosophical and even metaphysi-
cal reasoning, thereby confusing the issue even further.[3]

✧   ✧   ✧

There is hardly a branch of mathematics where geometric
progressions—finite or infinite—do not play a role. We first encounter them in arithmetic in the form of repeating decimals,
which are but infinite geometric progressions in disguise; for
example, the repeating decimal 0.1212. . . is merely an abbreviation for the infinite series $12/100 + 12/100^2 + 12/100^3 + \cdots$.
Geometric progressions are at the heart of most financial calculations, a result of the fact that money invested at a fixed
interest rate grows geometrically with time. In calculus we are
introduced to power series, and the simplest power series is the
infinite geometric progression $1 + x + x^2 + \cdots$, often used to
test the convergence of other series. Archimedes of Syracuse
(ca. 287–212 B.C.) cleverly used a geometric progression to find
the area of a parabolic segment—one of the first quadratures of
a curved shape.[4] And the fractals of modern vintage, those in-
tricate self-replicating curves that meander endlessly hither and
thither, are but an application of the principle of self-similarity,
of which the geometric progression is the simplest case (fig. 53).
The Dutch artist Maurits C. Escher (1898–1972), whose mathematical drawings have intrigued a whole generation of scientists,
used geometric progressions in several of his prints; we show
here one of them, entitled *Smaller and Smaller* (Fig. 54).

✧   ✧   ✧

A common misconception among students of mathematics (un-
doubtedly fueled by erroneous statements found in popular
books) is that Euclid's great work, the *Elements*, deals only with
geometry. True, geometry occupies the bulk of the work, but it
also contains an extensive treatment of arithmetic, number theory, and the theory of progressions. All of Book VIII and parts
of Book IX are devoted to "continued proportions," that is,
numbers that form a geometric progression (a favorite subject
with the Greeks ever since Pythagoras's discovery that musical
intervals correspond to simple proportions of string lengths).

FIG. 53. Construction of the snowflake curve: start with an equilateral triangle, construct a smaller equilateral triangle over the middle third of each side, and cut the middle third to obtain a Star of David–like figure. Repeat the process with the new figure to get a 48-sided figure. Continuing in this manner, we get a sequence of shapes that, in the limit, approach a crinkly curve known as the *snowflake curve* (also called the Koch curve after its discoverer, the Swedish mathematician Helge von Koch [1870–1924]). The perimeter and area of these shapes follow geometric progressions with common ratios 4/3 and 4/9, respectively. Since these ratios are, respectively, greater than and less than 1, the perimeter tends to infinity, while the area tends to 8/5 the area of the original triangle. The snowflake curve was the first known "pathological curve"; it is nowhere smooth and hence has nowhere a derivative. Today such self-replicating shapes are called *fractals*.

Proposition 35 of Book IX states in words how to find the sum of a geometric progression:

If as many numbers as we please be in continued proportion, and there be subtracted from the second [number] and the last [number] numbers equal to the first, then, as the excess of the second is to the first, so will the excess of the last be to all those before it.

Translated into modern language, if the terms of the progression are $a, ar, ar^2, \ldots, ar^n$ and the sum of "all those before it" is S, then $(ar - a) : a = (ar^n - a) : S$; cross-multiplying and simplifying, we get the familiar formula for the sum of the first n terms of a geometric progression,

$$S = \frac{a(r^n - 1)}{r - 1}.^5 \tag{1}$$

FIG. 54. M. C. Escher's *Smaller and Smaller* (1956). ©1997
Cordon-Art-Baarn-Holland. All rights reserved.

Euclid then uses this result to prove (Proposition 36 of
Book IX) an elegant property of numbers: if the sum of the
progression $1 + 2 + 2^2 + \cdots + 2^{n-1}$ is a prime number, then the
product of this prime and 2^{n-1} is a *perfect number*. A positive
integer N is perfect if it is the sum of its positive divisors other
than N; the first two perfect numbers are $6 = 1 + 2 + 3$ and
$28 = 1 + 2 + 4 + 7 + 14$. Since the sum of the progression $1 + 2 +
2^2 + \cdots + 2^{n-1}$ is $2^n - 1$, the proposition says that $2^{n-1} \cdot (2^n - 1)$
is perfect whenever $2^n - 1$ is prime. Thus 6 is perfect because
$6 = 2 \cdot 3 = 2^{2-1} \cdot (2^2 - 1)$ and 28 is perfect because $28 = 4 \cdot 7 =
2^{3-1} \cdot (2^3 - 1)$. The next two perfect numbers are $496 = 16 \cdot 31 =
2^{5-1} \cdot (2^5 - 1)$ and $8,128 = 64 \cdot 127 = 2^{7-1} \cdot (2^7 - 1)$. These four
were the only perfect numbers known to the Greeks.[6]

And this is as far as the Greeks went. They made effective use
of equation (1) in their development of geometry and number
theory, allowing n to be arbitrarily large ("as many numbers as
we please"); but they did not make the crucial step of actually

letting n grow beyond all bounds—letting it tend to infinity. Had they not limited themselves with this self-imposed taboo, they just might have anticipated the discovery of the calculus by two thousand years.[7]

Today, with the limit concept firmly established, we have no qualms arguing that if r is a number whose absolute value is less than 1 ($-1 < r < 1$), then as $n \to \infty$, the term r^n in equation (1) tends to zero, so that in the limit we get $S = -a/(r-1)$, or equivalently

$$S = \frac{a}{1-r}, \tag{2}$$

which is the familiar formula for the sum of an *infinite* geometric progression.[8] Thus the series in Zeno's paradox, $1/2 + 1/4 + 1/8 + 1/16 + \cdots$, has the sum $(1/2)/(1 - 1/2) = 1$, and the repeating decimal $0.1212 \cdots = 12/100 + 12/10,000 + \cdots$ has the sum $(12/100)/(1 - 1/100) = 12/99 = 4/33$. We can actually use equation (2) to prove that *every* repeating decimal is equal to some fraction, i.e., a rational number.

❖ ❖ ❖

And now trigonometry enters the picture. We will show that *every infinite geometric progression can be constructed geometrically, and its sum found graphically, using only a straightedge and compass.*[9] Our starting point is the fact that the sum of an infinite geometric progression with common ratio r converges to a limit if and only if $-1 < r < 1$. Now any number between -1 and 1 is the cosine of exactly one angle between $0°$ and $180°$; for example, 0.5 is the cosine of $60°$, and -0.707 (more precisely, $-\sqrt{2}/2$) is the cosine of $135°$. (Note that this is not so with the sine function: there are *two* angles, $30°$ and $(180° - 30°) = 150°$, whose sine is 0.5, and *no* angle between $0°$ and $180°$ whose sine is -0.707). Let us, therefore, write $r = \cos \alpha$, or conversely $\alpha = \cos^{-1} r$, and regard the angle α as given.

On the x-axis let the origin be at P_0 and the point $x = 1$ at P_1 (fig. 55). At P_1 we draw a ray forming the angle α with the positive x-axis, and along it mark a segment $P_1 Q_1$ of unit length. From Q_1 we drop a perpendicular to the x-axis, meeting it as P_2; we have $P_1 P_2 = 1 \cdot \cos \alpha = \cos \alpha$ and thus $P_0 P_2 = 1 + \cos \alpha$. We now repeat the process: at P_2 we draw a ray forming the angle α with the positive x-axis and along it mark a segment $P_2 Q_2$ equal in length to $P_1 P_2$ (using a compass centered at P_2 and opened to the length $P_1 P_2$). From Q_2 we drop a perpendicular to the x-axis, meeting it at P_3; we have $P_2 P_3 = \cos \alpha \cdot \cos \alpha = \cos^2 \alpha$ and thus $P_0 P_3 = 1 + \cos \alpha + \cos^2 \alpha$. Continuing in this manner, it seems

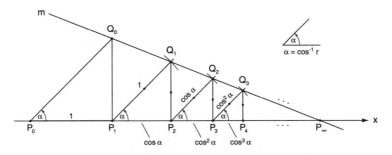

FIG. 55. Geometric construction of the series
$S = 1 + \cos\alpha + \cos^2\alpha + \cdots$.

at first that we would have to repeat the process infinitely many
times. But, as we will now show, *the first two steps are sufficient
to determine the sum of the entire series.*

First, the right triangles $P_1Q_1P_2$, $P_2Q_2P_3$, and so on are similar, having the same angle α; consequently the points $Q_1, Q_2, \cdots$
must lie on a straight line m. We claim that the point of intersection of m with the x-axis marks the sum S of the entire series, and we accordingly denote this point P_∞. To prove this, we
note that the segments $P_1Q_1 = 1$, $P_2Q_2 = \cos\alpha$, $P_3Q_3 = \cos^2\alpha$,
and so on form a geometric progression with the common ratio
$\cos\alpha$ (the same as in the original progression). Going one step
backward in this progression, we have $P_0Q_0 = 1/\cos\alpha = \sec\alpha$.
Now the oblique triangles $P_0Q_0P_\infty$, $P_1Q_1P_\infty$, $\cdots$ are all similar;
taking the first two of these triangles, we have $P_0P_\infty/P_0Q_0 = P_1P_\infty/P_1Q_1$, or

$$\frac{S}{\sec\alpha} = \frac{S-1}{1}.$$

Changing the factor $1/\sec\alpha$ back to $\cos\alpha$ and solving the equation for S, we get $S = 1/(1 - \cos\alpha) = 1/(1 - r)$, showing that
the segment P_0P_∞ is the sum of the entire series. We repeat: it
is sufficient to construct the first two points Q_1 and Q_2; these
determine the line m, whose intersection with the x-axis deter-
mines the point P_∞.

Not only does this construction provide a geometric interpretation of the geometric series, it also allows us to see what happens when we vary the common ratio r. Figures 56 and 57 show
the construction for $\alpha = 60°$ and $45°$, for which $r = 1/2$ and
$\sqrt{2}/2$, respectively; the corresponding sums are $1/(1 - 1/2) = 2$
and $1/(1 - \sqrt{2}/2) = 2 + \sqrt{2} \approx 3.414$. As we vary r and with it
α, points P_0 and P_1 remain fixed, but all other points will move
along their respective lines. For $\alpha = 90°$ (that is, $r = 0$), Q_1 will
be exactly above P_1, so that dropping the perpendicular from it

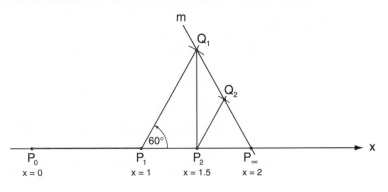

FIG. 56. The construction for $\alpha = 60°$.

to the x-axis brings us back to P_1: the series will not progress any further, and its sum is $S = P_0P_1 = 1$. As we decrease α from 90° to 0°, the line m becomes less and less steep; at the same time, points $P_2, P_3, \cdots$ move to the right, and so does P_∞: the sum of the series becomes larger. As $\alpha \to 0°$, the line m becomes horizontal, and its point of intersection with the x-axis recedes to infinity: the series diverges.

If the common ratio r is negative, α will be between 90° and 180°. Starting again at P_1 (fig. 58), we draw a ray making with the positive x-axis the (obtuse) angle α; this brings us to point Q_1, with $P_1Q_1 = 1$. We now drop a perpendicular from Q_1 to the x-axis, meeting it at P_2 (note that P_2 is now to the *left* of P_1); we have $P_1P_2 = \cos\alpha$ (a negative number) and thus $P_0P_2 = 1 + \cos\alpha$. From P_2 we draw a ray forming the angle α with the positive x-axis; note that since segment P_1P_2 is directed to the left, the ray will be directed *downward*. On this ray we mark off a segment P_2Q_2 equal in length to P_1P_2. From Q_2 we drop a perpendicular to the x-axis, meeting it at P_3 (note that P_3 is to the *right* of P_2); we have $P_2P_3 = \cos^2\alpha$ (a positive number), hence $P_0P_3 = 1 + \cos\alpha + \cos^2\alpha$. Proceeding in this way,

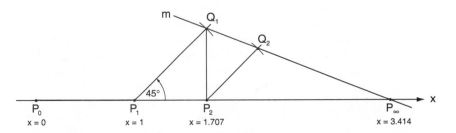

FIG. 57. The construction for $\alpha = 45°$.

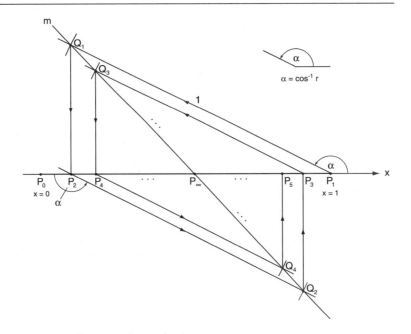

FIG. 58. The case when α is obtuse.

we get ever smaller right triangles, each nested within the one preceding it by two steps. All these triangles are similar.

It follows, as before, that the points $Q_1, Q_2, \cdots$ lie on a straight line m, whose point of intersection with the x-axis gives us the sum of the entire series. Denoting this point by P_∞, we note that it lies to the right of the points P_{2n} and to the left of the points P_{2n+1}: the series approaches its sum alternately from above and below, depending on whether we have summed up an odd or an even number of terms. Figures 59 and 60 show the construction for $\alpha = 120°$ and $150°$ (that is, $r = -1/2$ and $-\sqrt{3}/2$, respectively), for which the series converges to $1/(1 + 1/2) = 2/3 \approx 0.666$ and $1/(1 + \sqrt{3}/2) \approx 0.536$.

Now let us once again vary the angle α, this time increasing it from 90° to 180°. The line m will assume a less and less steep position, with, however, a negative slope. At the same time the points P_{2n} will move to the left toward P_0, while the points P_{2n+1} will move to the right toward P_1. As $\alpha \to 180°$ (that is, as $r \to -1$), the points Q_{2n+1} will cluster above P_0 and the points Q_{2n} will cluster below P_1, so that the line m will assume a nearly symmetrical position relative to the segment P_0P_1, intersecting it just to the right of the point $x = 1/2$. And this, indeed, is the value to which the formula $S = 1/(1 - r)$ tends as $r \to -1$.

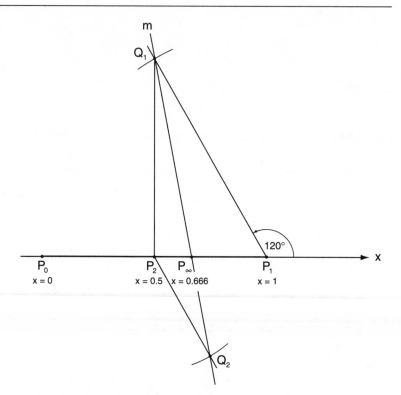

FIG. 59. The construction for $\alpha = 120°$.

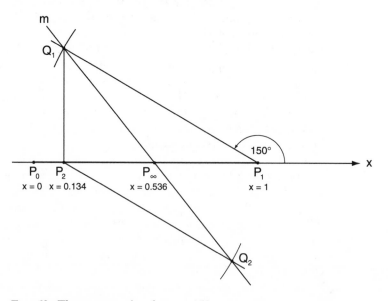

FIG. 60. The construction for $\alpha = 150°$.

At the same time, however, the points P_{2n} will crowd about P_0 (that is, about $x = 0$), while the points P_{2n+1} will crowd about P_1 ($x = 1$), showing that the series will tend to oscillate between 0 and 1.

When $\alpha = 180°$ (that is, $r = -1$) the situation suddenly changes, for then all the points Q_{2n+1} merge with P_0 (as also the points P_{2n}), while the points Q_{2n} merge with P_1. The line m will then merge with the x-axis, intersecting it at infinitely many points, and it becomes impossible to determine the point P_∞. At first this situation seems contradictory, since for $\alpha = 180°$ our series becomes $1 - 1 + 1 - 1 + - \cdots$, whose partial sums oscillate between 0 and 1. But this is not so! The series can actually be made to equal any arbitrary number, which merely shows that the series does not converge and its sum has no meaning.[10]

The seemingly bizarre behavior of the series $1 - 1 + 1 - 1 + - \cdots$ aroused much controversy in the early eighteenth century. Gottfried Wilhelm Leibniz (1646–1716), co-inventor with Newton of the calculus, argued that since the sum may be 0 or 1 with equal probability, its "true" value should be their mean, namely 1/2, in agreement with the formula $S = 1/(1 - r)$ when $r = -1$. Such careless reasoning may seem incredible to us today, but in Leibniz's time the concepts of convergence and limit were not yet understood, and infinite series were treated in a purely manipulative manner, as if they were an extension of ordinary finite sums.

As he groped with this series, Leibniz—who was a philosopher at heart—must have thought of Zeno, his predecessor by two thousand years. Had Zeno been aware of our construction, perhaps it might have made it easier for him to accept the fact that an infinite sum of numbers may be finite. And the consequences would have been profound, for had the Greeks not been so stubborn in barring infinity from their world, the course of mathematics might have been forever changed.

NOTES AND SOURCES

1. A variation of the paradox says that for the runner to go from A to B he must first reach the midpoint C between A and B; but in order to reach C, he must first reach the midpoint D between A and C, and so on.

2. For the causes of this fear, see my book, *e: The Story of a Number* (Princeton, N.J.: Princeton University Press, 1994), pp. 43–47.

3. Even today some thinkers refuse to regard Zeno's paradoxes as settled; see the articles "Resolving Zeno's Paradoxes" by William I. McLaughlin, *Scientific American*, November 1994, and "A Brief History of Infinity" by A. W. Moore, *Scientific American*, April 1995. See also Adolf Grünbaum, *Modern Science and Zeno's Paradoxes* (Middletown, Conn.: Wesleyan University Press, 1967).

4. See the chapter "Quadrature of the Parabola" in Thomas L. Heath, *The Works of Archimedes* (1897; rpt. New York: Dover, 1953).

5. A modern proof is to write $S = a + ar + ar^2 + \cdots + ar^{n-1}$, multiply this equation by r, and subtract the result from the original equation: all terms except the first and last will cancel, giving us $(1 - r)S = a - ar^n$, from which we get $S = a(1 - r^n)/(1 - r) = a(r^n - 1)/(r - 1)$.

6. Note that $2^n - 1$ is not prime for *every* prime n; for example, $2^{11} - 1 = 2047 = 23 \cdot 89$ is composite, and therefore $2^{11-1} \cdot (2^{11} - 1) = 2{,}096{,}128$ is not perfect. Primes of the form $2^n - 1$ where n is prime are called *Mersenne primes*, named after Marin Mersenne (1588–1648), the French friar of the order of Minims; as of 1996 only thirty-four Mersenne primes were known, the largest being $2^{1{,}257{,}787} - 1$, a 378,632-digit number discovered in that year. Because every Mersenne prime generates a perfect number, their histories are closely related.

By necessity, the formula $2^{n-1} \cdot (2^n - 1)$ produces only *even* perfect numbers. In 1770 Leonhard Euler proved the converse of Proposition 36: *Every* even perfect number must be of the form $2^{n-1} \cdot (2^n - 1)$, where $2^n - 1$ is prime. It is not known whether odd perfect numbers exist, nor whether the number of perfect numbers is finite or infinite. For further details, see any good book on number theory.

7. See Heath, *Works of Archimedes*, chap. 7, "Anticipations by Archimedes of the Integral Calculus".

8. A common (though not quite rigorous) proof of equation (2) is to write $S = a + ar + ar^2 + ar^3 + \cdots = a + r(a + ar + ar^2 + \cdots) = a + rS$, from which we get $S(1 - r) = a$ or $S = a/(1 - r)$.

9. The subsequent material is based on my article "Geometric Construction of the Geometric Series" in the *International Journal of Mathematics Education in Science and Technology*, vol. 8, no. 1 (January 1977), pp. 89–96.

10. To show this, let a and b be any two numbers such that $a + b = 1$. Our series then becomes $(a + b) - (a + b) + (a + b) - (a + b) + - \cdots$. Let us call its sum S. Shifting the parentheses one position to the right, we obtain the series $S = a + (b - a) - (b - a) + (b - a) - (b - a) + - \cdots$. Now put $b - a = c$. Then $S = a + c - c + c - c + - \cdots$. We can sum this last series in two ways, depending on how we arrange the parentheses: $S = a + (c - c) + (c - c) + (c - c) + \cdots = a$, or $S = a + c - (c - c) - (c - c) - (c - c) - \cdots = a + c = a + (b - a) = b$. Thus the series can have either a or b as its sum, and as the splitting of 1 into a and b was entirely arbitrary, S can have any value whatsoever. This, of course, merely shows that the partial sums do not converge to a fixed value, and thus the series diverges (though not to infinity).

10

(sin x)/x

Students of calculus encounter the function $(\sin x)/x$ early in their study, when it is shown that $\lim_{x \to 0}(\sin x)/x = 1$; this result is then used to establish the differentiation formulas $(\sin x)' = \cos x$ and $(\cos x)' = -\sin x$. Once this has been done, however, the function is soon forgotten, and the student rarely sees it again. This is unfortunate, for this simple-looking function not only has some remarkable properties, but it also shows up in many applications, sometimes quite unexpectedly.

We note, to begin with, that the function is defined for all values of x except 0; but we also know that as x gets smaller and smaller, the ratio $(\sin x)/x$—provided x is measured in radians—tends to 1. This provides us with a simple example of a *removable singularity*: we can simply *define* the value of $(\sin 0)/0$ to be 1, and this definition will assure the continuity of the function near $x = 0$.

Let us denote our function by $f(x)$ and plot it for various values of x; the resulting graph is shown in figure 61. Two features make this graph distinct from that of the function $g(x) = \sin x$: first, it is symmetric about the y-axis; that is, $f(-x) = f(x)$ for all values of x (in the language of algebra, $f(x)$ is an *even function*, so called because the simplest functions with this property are of the form $y = x^n$ for even values of n). By contrast, the function $g(x) = \sin x$ has the property that $g(-x) = -g(x)$ for all x (functions with this property are called *odd functions*, for example $y = x^n$ for odd values of n). To prove that $f(x) = (\sin x)/x$ is even, we simply note that $f(-x) = \sin(-x)/(-x) = (-\sin x)/(-x) = (\sin x)/x = f(x)$.

Second, unlike the graph of $\sin x$, whose up-and-down oscillations are confined to the range from -1 to 1 (that is, the sine

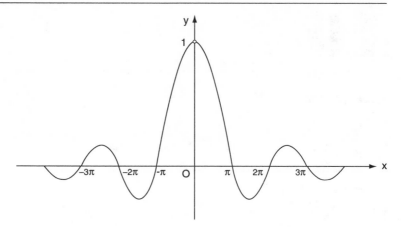

FIG. 61. The graph of $(\sin x)/x$.

wave has a constant amplitude 1), the graph of $(\sin x)/x$ represents *damped* oscillations whose amplitude steadily decreases as $|x|$ increases. Indeed, we may think of $f(x)$ as a sine wave squeezed between the two *envelopes* $y = \pm 1/x$. We now wish to locate the *extreme points* of $f(x)$—the points where it assumes its maximum or minimum values. And here a surprise is awaiting us. We know that the extreme points of $g(x) = \sin x$ occur at all odd multiples of $\pi/2$, that is, at $x = (2n+1)\pi/2$. So we might expect the same to be true for the extreme points of $f(x) = (\sin x)/x$. This, however, is not the case. To find the extreme point, we differentiate $f(x)$ using the quotient rule and equate the result to zero:

$$f'(x) = \frac{x \cos x - \sin x}{x^2} = 0. \tag{1}$$

Now if a ratio is equal to zero, then the numerator itself must equal to zero, so we have $x \cos x - \sin x = 0$, from which we get

$$\tan x = x. \tag{2}$$

Equation (2) cannot, unfortunately, be solved by a closed formula in the same manner as, say, a quadratic equation can; it is a *transcendental equation* whose roots can be found graphically as the points of intersection of the graphs of $y = x$ and $y = \tan x$ (fig. 62). We see that there is an infinite number of these points, whose x-coordinates we will denote by x_n. As x increases in absolute value, these points rapidly approach the asymptotes of $\tan x$, that is, $(2n+1)\pi/2$; these, of course, are the extreme points of $\sin x$. This is to be expected, since as $|x|$ increases, $1/|x|$ decreases at a rate that itself is decreasing, so

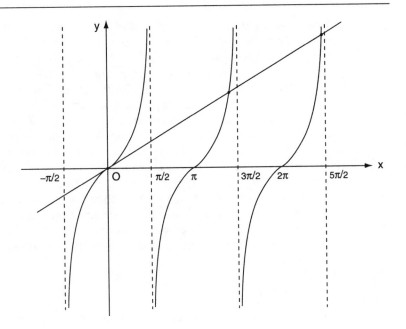

FIG. 62. The roots of $\tan x = x$.

that its effect on the variation of $\sin x$ steadily diminishes. The first few values of x_n are given in table 2.

The peculiar behavior of the extreme points of $(\sin x)/x$ is in marked contrast to another kind of damped oscillations, those represented by the function $e^{-x} \sin x$. Here the extreme points are shifted leftward by a *constant* amount $\pi/4$ relative to those of $\sin x$, as the reader can easily verify.

Having explored the general shape of the graph of $f(x)$, the next question of interest is to find the area under the graph from, say, $x = 0$ to some other x. This area is given by the definite integral

$$\int_0^x \frac{\sin t}{t} dt.$$

TABLE 2

n	x_n	$f(x_n)$
0	0.00	1.000
1	$4.49 = 2.86\pi/2$	-0.217
2	$7.73 = 4.92\pi/2$	0.128
3	$10.90 = 6.94\pi/2$	-0.091
4	$14.07 = 8.96\pi/2$	0.071

where we have denoted the variable of integration by t to distinguish it from the upper limit x. In order to evaluate this integral, we would first seek to find the indefinite integral, or antiderivative, of $(\sin x)/x$. Alas, this is a futile attempt! It is one of the curious facts of calculus that the antiderivatives of many simple-looking functions cannot be expressed in terms of the "elementary functions," that is, polynomials and ratios of polynomials, exponential and trigonometric functions and their inverses, and any finite combination of these functions. The function $(\sin x)/x$ belongs to this group, as do $(\cos x)/x$, e^x/x, and e^{x^2}. This, of course, does not mean that the antiderivatives of these functions do not exist—it only means that they cannot be expressed in "closed form" in terms of the elementary functions. Indeed, the above integral, regarded as a function of its upper limit x, *defines* a new, "higher" function known as the *sine integral* and denoted by $\operatorname{Si}(x)$:

$$\operatorname{Si}(x) = \int_0^x \frac{\sin t}{t}\,dt.$$

Although we cannot express $\operatorname{Si}(x)$ in terms of the elementary functions, we can nevertheless compute its values and plot them on a graph (fig. 63). This is done by writing the sine function as a power series, $\sin x = x - x^3/3! + x^5/5! - + \cdots$, dividing each term by x, and then integrating term by term. The result is

$$\operatorname{Si}(x) = x - x^3/3 \cdot 3! + x^5/5 \cdot 5! - + \cdots,$$

a series that converges for all x.

As we let the upper limit x increase without bound, will the area under the graph approach a limit? The answer is yes; it can be shown that this limit is $\pi/2$;[1] in other words,

$$\operatorname{Si}(\infty) = \int_0^\infty \frac{\sin x}{x}\,dx = \pi/2. \tag{3}$$

This important integral is known as the *Dirichlet integral*, after the German mathematician Peter Gustav Lejeune Dirichlet (1805–1859). An unexpected by-product of this integral is obtained by replacing $\sin x$ with $\sin kx$, where k is a constant, and then making the substitution $u = kx$. We find that the new integral has the value $\pi/2$ or $-\pi/2$, depending on whether k is positive or negative (in the latter case the upper limit becomes $-\infty$, so that the further substitution $v = -u$ will result in $-\pi/2$). We thus have the following result:

$$(2/\pi) \int_0^\infty \frac{\sin kx}{x}\,dx = \begin{cases} 1 & \text{for } k > 0 \\ 0 & \text{for } k = 0 \\ -1 & \text{for } k < 0 \end{cases}. \tag{4}$$

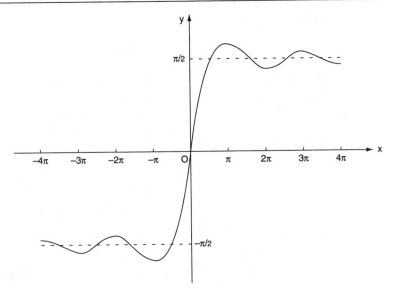

FIG. 63. The graph of Si $(x) = \int_0^x (\sin t)/t\, dt$.

But the expression on the right, considered as a function of k, is the "sign function" shown in figure 64. We have here one of the simplest examples of an integral representation of a function; the need for such a representation often arises in applied mathematics. The integral on the left is known as *Dirichlet's discontinuity factor*.

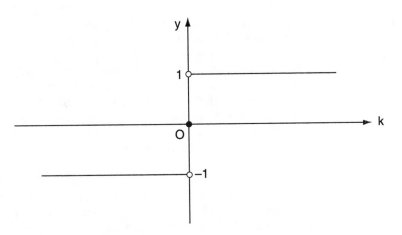

FIG. 64. The graph of sign x.

✧ ✧ ✧

Of the many occurrences of the function $(\sin x)/x$, we will consider here one taken from geography. Early in school we learn that our earth is round, though it took many centuries before this fact was universally accepted (the last of the flat-earth believers finally gave up when images from spacecraft showed the earth to be round). Indeed, to the uninitiated it is not at all immediately obvious that we live on a round world—certainly most of our daily experiences could more naturally be explained on the basis of a flat earth. It is only indirectly, chiefly through astronomical observations, that we know the earth is round.

In his classic mathematical novel *Flatland*, Edwin A. Abbott describes the world of two-dimensional, antlike creatures who can move forward and backward and left and right, but not up and down. If these "flatlanders" were to inhabit our earth, they would be unaware of its sphericity: from their viewpoint the earth would seem as flat as a tabletop. But one day they decide to explore their world, intent on discovering its underlying geometry. Starting at the North Pole and using a stretched rope as a compass, they draw circles around the pole with ever larger radii. Then they measure the circumference of each circle and express it in terms of its radius. Back at home, they put to a test what they had learned in school—that the ratio of the circumference of a circle to its radius is the same for all circles, about 6.28. For small circles they find, to their delight, that this indeed seems to be the case. But as the circles get larger, reassurance turns into doubt, then disappointment: our flatlanders find that the circumference-to-radius ratio is not constant after all.

To see the reason for this, let us take advantage of the privilege granted to us, humans, by being three-dimensional creatures: *we* know that our world is round. Let us denote the radius of the earth—assumed to be a perfect sphere—by R. To find the circumference of a circle around the North Pole, we need to know its radius, and this depends on the geographic latitude of the circle. If, for simplicity, we measure the latitude not from the equator, as is done in geography, but from the North Pole, then the radius of a circle of latitude θ is $r = R \sin \theta$ (see fig. 65), and its circumference is

$$c = 2\pi R \sin \theta. \tag{5}$$

This result, of course, is entirely satisfactory to us three-dimensionals, but to our two-dimensional earth dwellers it is totally meaningless. They have no idea that they live on a curved surface, and if someone told them that their flat

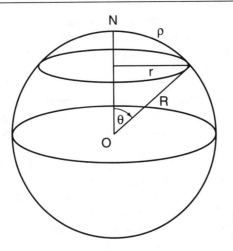

FIG. 65. Circle of latitude θ on the globe.

world is actually spherical, they would be puzzled indeed. For them a quantity such as R, taken from the third dimension and not being capable of direct measurement, is as meaning-less as for an elementary school pupil to find the volume of a four-dimensional "sphere."

To make the formula meaningful, we must express it in terms of variables that our inhabitants can measure. Indeed, the most important variable, from their point of view, is the radius of the circle, *as measured on the surface of the earth*. Let us denote this radius by the Greek letter ρ (rho). If we measure θ in radians, we have $\rho = R\theta$, hence $R = \rho/\theta$. Substituting this expression in equation (5), we get

$$c = 2\pi\rho\frac{\sin\theta}{\theta}. \tag{6}$$

Thus the circumference depends not only on the radius, but also on the latitude.

Before contemplating the consequences of this formula, we may wonder how our inhabitants would measure the latitude θ when they are unaware of the sphericity of their world. They might get a clue by watching the sky above them: they might notice, as mariners in ancient times did, that the entire celestial sphere appears to rotate once every 24 hours around one star that seems to be standing still—the North Star. Moreover, the height of the North Star above the horizon steadily decreases as they travel southward; in fact, they find that the angle θ between the North Star and the zenith—the point on the celestial sphere directly above the observer—is proportional to the distance ρ from the North Pole (as follows from the equation $\rho = R\theta$).

And now our inhabitants are ready to put to a test what they had learned in their geometry class. For small latitudes (angular distances from the North Pole), they will find that the ratio c/ρ does indeed appear to be constant, or nearly so, as table 3 shows. Their surveyors might at first dismiss the small discrepancies from constancy as due to errors of measurement, but it will soon become clear that the ratio c/ρ is *not* constant but decreases with θ, as table 4 shows. (The 4 in the last entry reflects the fact that the distance from the pole to the equator is exactly one-fourth the circumference of the equator.) Had our inhabitants extended the table further—that is, into the southern hemisphere—the ratio c/ρ would continue to decrease, until it becomes zero at 180°(the South Pole). Still unaware that their world is round, they will have lost any remaining faith in what they had learned about the constancy of the circumference-to-radius ratio.[2] But perhaps some wise flatlander might interpret these findings differently and conclude that the world they live in is actually curved. That wise flatlander would go down in history as the discoverer of the third dimension.

We can actually draw a map of the world as the flatlanders would see it. Known as an azimuthal equidistant map, it shows all "straight line" distances and directions from a fixed, preselected point, located at the center of the map, to any other point on the globe. (A "straight line" between two points on a sphere is an arc of the *great circle* connecting them—the circle passing through the two points and having its center at the center of the sphere [fig. 66]; it represents the shortest distance between the points.) Figure 67 shows such a map centered on San Francisco; we see that the direct route from San Francisco to Moscow passes over the North Pole, and that Moscow is closer

TABLE 3

θ	c/ρ
0°	6.283
1°	6.283
2°	6.282
3°	6.280
4°	6.278
5°	6.275

Note: in using equation (6), all angles must first be converted to radians ($1° = \pi/180$ radians).

TABLE 4

θ	c/ρ
10°	6.251
20°	6.156
30°	6.000
40°	5.785
50°	5.516
60°	5.196
70°	4.833
80°	4.432
90°	4.000

to San Francisco than San Francisco is to Rio de Janeiro. Notice that Africa, Antarctica, and Australia appear extremely distorted, both in shape and size; this is because a circle of radius ρ centered at the fixed point has circumference $2\pi\rho$ on the map, while on the globe its circumference is $2\pi\rho(\sin\theta)/\theta$, where θ has the same meaning as in figure 65, but with the fixed point replacing the North Pole. Concentric circles around the fixed point are thus exaggerated by the ratio $1: [(\sin\theta)/\theta]$, or $\theta/\sin\theta$, relative to their true size on the globe. This exaggeration factor increases with θ and becomes infinite when $\theta = 180°$, that is, at *the antipode* of the fixed point (the point opposite to it on the globe). On an azimuthal equidistant map, the entire outer boundary represents the antipode of the central point; it marks the edge of the universe of our flatlanders—the farthest point they can reach in any direction. They discover that their universe, though unbounded, is finite.

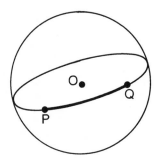

FIG. 66. Arc of a great circle.

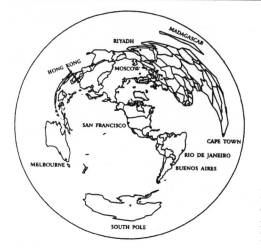

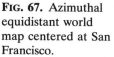

FIG. 67. Azimuthal equidistant world map centered at San Francisco.

NOTES AND SOURCES

1. The proof is not elementary; see Richard Courant, *Differential and Integral Calculus* (London: Blackie & Son, 1956), vol. 1, pp. 251–253 and 444–450; for an alternative proof using integration in the complex plane, see Erwin Kreyszig, *Advanced Engineering Mathematics* (New York: John Wiley, 1979), pp. 735–736.

2. A similar situation arises in connection with the area of a "circle" of radius ρ (actually a spherical cap). This area is given by $A = 2\pi Rh$, where h is the height of the cap (the distance from its base to the surface of the sphere). Now $h = R(1 - \cos\theta) = 2R\sin^2(\theta/2)$, so that $A = 4\pi R^2 \sin^2(\theta/2) = 4\pi(\rho/\theta)^2 \sin^2(\theta/2) = \pi\rho^2\{[\sin(\theta/2)]/(\theta/2)\}^2$. A "correction factor" of $\{[\sin(\theta/2)]/(\theta/2)\}^2$ is thus needed if we want to find the ratio A/ρ^2.

11

A Remarkable Formula

The prototype of all infinite processes is repetition. . . .
Our very concept of the infinite derives from the notion
that what has been said or done once can always be
repeated.
—Tobias Dantzig, *Number: The Language of Science*

We are not quite done yet with the function $(\sin x)/x$. Browsing one day through a handbook of mathematical formulas, I came across the following equation:

$$\frac{\sin x}{x} = \cos \frac{x}{2} \cdot \cos \frac{x}{4} \cdot \cos \frac{x}{8} \cdots . \tag{1}$$

As I had never seen this formula before, I expected the proof to be rather difficult. To my surprise, it turned out to be extremely simple:

$$\sin x = 2 \sin x/2 \cdot \cos x/2$$

$$= 4 \sin x/4 \cdot \cos x/4 \cdot \cos x/2$$

$$= 8 \sin x/8 \cdot \cos x/8 \cdot \cos x/4 \cdot \cos x/2$$

$$= \cdots$$

After repeating this process n times, we get

$$\sin x = 2^n \sin x/2^n \cdot \cos x/2^n \cdot \ldots \cdot \cos x/2.$$

Let us multiply and divide the first term of this product by x (assuming, of course, that $x \neq 0$) and rewrite it as $x \cdot [(\sin x/2^n)/(x/2^n)]$; we then have

$$\sin x = x \cdot \left[\frac{\sin x/2^n}{x/2^n} \right] \cdot \cos x/2 \cdot \cos x/4 \cdot \ldots \cdot \cos x/2^n.$$

Note that we have reversed the order of the remaining terms in the (as yet finite) product. If we now let $n \to \infty$ while keeping

x constant, then $x/2^n \to 0$ and the expression in the brackets, being of the form $(\sin\alpha)/\alpha$, will tend to 1. We thus have

$$\sin x = x \prod_{n=1}^{\infty} \cos x/2^n,$$

where $\prod$ stands for "product." Dividing both sides by x, we get equation (1).

Equation (1) was discovered by Euler[1] and represents one of the very few examples of an *infinite product* in elementary mathematics. Since the equation holds for all values of x (including $x = 0$, if we define $(\sin 0)/0$ to be 1), we can substitute in it, for example, $x = \pi/2$:

$$\frac{\sin\pi/2}{\pi/2} = \cos\pi/4 \cdot \cos\pi/8 \cdot \cos\pi/16 \cdot \dots .$$

Now $\sin\pi/2 = 1$ and $\cos\pi/4 = (\sqrt{2})/2$. Using the half-angle formula $\cos x/2 = \sqrt{(1+\cos x)/2}$ for each of the remaining terms, we get, after a slight simplification,

$$\frac{2}{\pi} = \frac{\sqrt{2}}{2} \cdot \frac{\sqrt{2+\sqrt{2}}}{2} \cdot \frac{\sqrt{2+\sqrt{2+\sqrt{2}}}}{2} \cdot \dots .$$

This beautiful formula was discovered by Viète in 1593; in establishing it he used a geometric argument based on the ratio of areas of regular polygons of n and $2n$ sides inscribed in the same circle.[2] Viète's formula marks a milestone in the history of mathematics: it was the first time an infinite process was explicitly written as a succession of algebraic operations. (Up until then mathematicians were careful to avoid any direct reference to infinite processes, regarding them instead as a finite succession of operations that could be repeated as many times as one wished.) By adding the three dots at the end of his product, Viète, in one bold stroke, declared the infinite a bona fide part of mathematics. This marked the beginning of mathematical analysis in the modern sense of the word.

Aside from its beauty, Viète's formula is remarkable because it allows us to find the number π by repeatedly using four of the basic operations of arithmetic—addition, multiplication, division, and square root extraction—all applied to the number 2. This can be done even on the simplest scientific calculator:

$$2 \ \sqrt{x} \ \text{STO} \ \div \ 4 \ \times \ 2 \ \text{SUM} \ \text{RCL} \ \sqrt{x} \ \text{STO} \ \div \ 2$$

(on some calculators the memory operations STO, RCL, and SUM are labeled M, RM and M+, respectively). At each iteration you can read the current approximation of π by pressing the

$1/x$ key immediately after $\times$ in the key sequence shown above; then press $1/x$ again to start the next iteration. It is fascinating to watch the numbers in the display gradually approach the value of π; after the ninth iteration we get 3.1415914—a value correct to five places. A programmable calculator, of course, will speed up things considerably.

It is instructive to examine equation (1) from the point of view of convergence. We note, first of all, that the convergence of the partial products to their limiting value is *monotonic*; that is, each additional term carries us closer to the limit. This is because each term is a number less than 1, causing the value of the partial products to continually diminish. This is in marked contrast to the infinite *series* for $(\sin x)/x$,

$$\frac{\sin x}{x} = 1 - \frac{x^2}{3!} + \frac{x^4}{5!} - \frac{x^6}{7!} + - \cdots, \tag{2}$$

which approaches its limit alternately from above and below. The convergence, moreover, is very fast, although it is somewhat slower than that of the series. Table 5 compares the rates of convergence of equations (1) and (2) for $x = \pi/2$:

TABLE 5

Infinite Series	Infinite Product
$S_1 = 1.0000$	$\prod_1 = 0.7071$
$S_2 = 0.5888$	$\prod_2 = 0.6533$
$S_3 = 0.6395$	$\prod_3 = 0.6407$
$S_4 = 0.6365$	$\prod_4 = 0.6376$
$S_5 = 0.6366$	$\prod_5 = 0.6369$
$S_6 = 0.6366$	$\prod_6 = 0.6367$
$\cdots$	$\cdots$
$S_\infty = 0.6366$	$\prod_\infty = 0.6366$

Note: All figures are rounded to four decimal places.

The reason for the rapid convergence of the infinite product can be seen from figure 68. On the unit circle we mark off the radii corresponding to the angles $\theta/2, (\theta/2 + \theta/4), (\theta/2 + \theta/4 + \theta/8)$, and so on. These angles form an infinite geometric progression whose sum is $\theta/2 + \theta/4 + \theta/8 + \cdots = \theta$. Now, beginning at the x-axis, we take the perpendicular projection of each radius on the next radius. The lengths of these projections are 1, $\cos \theta/2$, $\cos \theta/2 \cdot \cos \theta/4$, $\cos \theta/2 \cdot \cos \theta/4 \cdot \cos \theta/8$, and so on. We see that after only a few steps the projections become barely distinguishable from their final value.

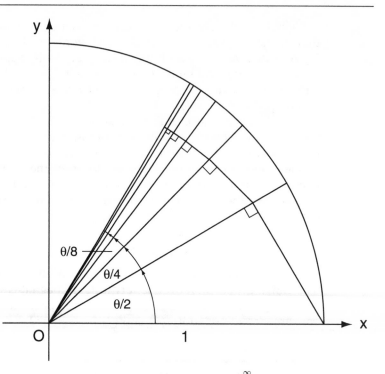

FIG. 68. Convergence of the infinite product $\prod\limits_{n=1}^{\infty} \cos x/2^n$.

Following the principle that every trigonometric identity can be interpreted geometrically, we now ask what geometric meaning we can give to equation (1). The answer is given in figure 69. We start with a circle of radius r_0 centered at the origin. From the x-axis we mark off an angle θ, whose terminal side intercepts the circle at the point P_0. We now connect P_0 with the point P_1 whose coordinates are $(-r_0, 0)$ and denote the segment $P_1 P_0$ by r_1. Angle OP_1P_0, having its vertex on the circle through P_1 and subtending the same arc as θ, is equal to $\theta/2$. Applying the Law of Sines to the triangle OP_1P_0, we have

$$\frac{r_0}{\sin(\theta/2)} = \frac{r_1}{\sin(180° - \theta)}. \tag{3}$$

But $\sin(180° - \theta) = \sin\theta = 2\sin\theta/2 \cdot \cos\theta/2$; putting this back in equation (3) and solving for r_1, we get $r_1 = 2r_0\cos\theta/2$.

We now draw a second circle, having P_1 as center and r_1 as radius. We have $\angle OP_2P_0 = \theta/4$, so that a repetition of the steps just performed and applied to the triangle $P_1P_2P_0$ gives us $r_2 = 2r_1\cos\theta/4 = 4r_0\cos\theta/2 \cdot \cos\theta/4$, where $r_2 = P_2P_0$. Repeating this process n times, we get a circle with center at P_n and radius

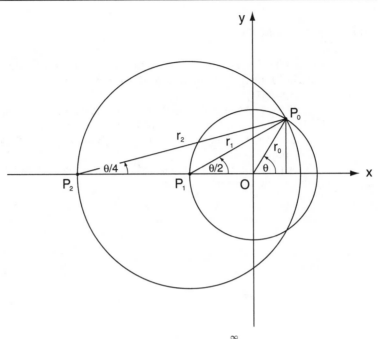

FIG. 69. Geometric proof of the formula $\prod_{n=1}^{\infty} \cos x/2^n = (\sin x)/x$.

$r_n = P_n P_0$ given by

$$r_n = 2^n r_0 \cos \theta/2 \cdot \cos \theta/4 \cdot \ldots \cdot \cos \theta/2^n. \tag{4}$$

Now $\angle OP_n P_0 = \theta/2^n$, so that the Law of Sines applied to the triangle $OP_n P_0$ gives us $r_0/(\sin \theta/2^n) = r_n/\sin(180° - \theta)$; hence

$$r_n = \frac{r_0 \sin \theta}{\sin \theta/2^n}. \tag{5}$$

Eliminating r_0 and r_n between equations (4) and (5), we get

$$(\sin \theta)/(\sin \theta/2^n) = 2^n \cos \theta/2 \cdot \cos \theta/4 \cdot \cdots \cdot \cos \theta/2^n. \tag{6}$$

As n increases without bound, angle $\angle OP_n P_0$ approaches zero and thus becomes indistinguishable from its sine; in other words, the arc whose radius is r_n tends to the perpendicular from P_0 to the x-axis. Replacing, then, $\sin \theta/2^n$ by $\theta/2^n$ and canceling the factor 2^n, we get equation (1), with θ instead of x.

Thus equation (1) is the trigonometric manifestation of the theorem that an angle inscribed in a circle has the same measure as one-half the central angle subtending the same arc, repeated again and again for ever smaller angles inscribed in ever larger circles.[3]

NOTES AND SOURCES

(This chapter is based on my article, "A Remarkable Trigonometric Identity," *Mathematics Teacher*, vol. 70, no. 5 (May 1977), pp. 452–455.)

1. E. W. Hobson, *Squaring the Circle: A History of the Problem* (Cambridge, England: Cambridge University Press, 1913), p. 26.

2. See Petr Beckmann, *A History of π* (Boulder, Colo.: Golem Press, 1977), pp. 92–96.

3. A proof of equation (1) based on physical considerations is given in the article mentioned above.

Jules Lissajous and His Figures

Jules Antoine Lissajous (1822–1880) is not among the giants in the history of science, yet his name is known to physics students through the "Lissajous figures"—patterns formed when two vibrations along perpendicular lines are superimposed. Lissajous entered the École Normale Supérieure in 1841 and later became professor of physics at the Lycée Saint-Louis in Paris, where he studied vibrations and sound. In 1855 he devised a simple optical method for studying compound vibrations: he attached a small mirror to each of the vibrating objects (two tuning forks, for example) and aimed a beam of light at one of the mirrors. The beam was reflected first to the other mirror and thence to a large screen, where it formed a two-dimensional pattern, the visual result of combining the two vibrations. This simple idea—a forerunner of the modern oscilloscope—was a novelty in Lissajous' time, for up until then the study of sound depended solely on the process of hearing, that is, on the human ear. Lissajous literally made it possible to "see sound."

✧   ✧   ✧

Let each vibration be a simple harmonic motion represented by a sinusoidal wave; let a and b denote the amplitudes, ω_1 and ω_2 the angular frequencies (in radians per second), ϕ_1 and ϕ_2 the phases, and t the time. We then have

$$x = a \sin (\omega_1 t + \phi_1), \, y = b \sin (\omega_2 t + \phi_2). \tag{1}$$

As time progresses, the point P whose coordinates are (x, y) will trace a curve whose equation can be found by eliminating t between equations (1). Since the two equations contain six parameters,[1] the curve is usually quite complicated, except in some special cases. For example, if $\omega_1 = \omega_2$ and $\phi_1 = \phi_1$, we have

$$x = a \sin (\omega t + \phi), \, y = b \sin (\omega t + \phi),$$

where we have dropped the subscripts under the parameters. To eliminate t between these equations, we note that $x/a = y/b$ and thus $y = (b/a)x$, the equation of a straight line. Similarly, for $\omega_1 = \omega_2$ and a phase difference of π we get the line $y =$

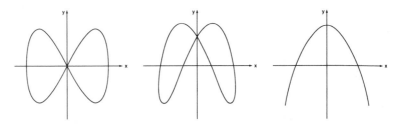

FIG. 71. Lissajous figures: the case $\omega_2 = 2\omega_1$.

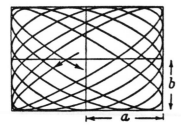

FIG. 72. Lissajous figure for the case when ω_1/ω_2 is irrational.

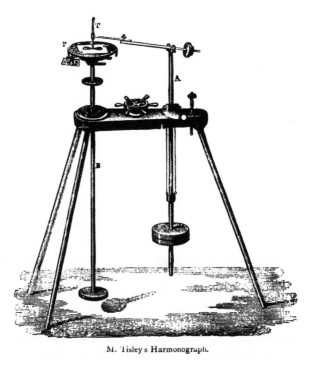

M. Tisley's Harmonograph.

FIG. 73. Harmonograph. From a nineteenth-century science book.

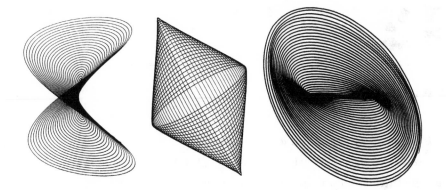

FIG. 74. Harmonograms.

ensuing figures were called "harmonograms," and their incredible variety never failed to impress the spectators (fig. 74).[4] The novelty of Lissajous' method was that it departed from mechanical devices and relied instead on the much more efficient agent of light. In this he was a visionary, foretelling our modern electronic era.

NOTES AND SOURCES

1. Actually five parameters are sufficient, since only the *relative* phase (i.e., the phase difference) matters.

2. The period will be the least common multiple of the two individual periods.

3. Florian Cajori, *A History of Physics* (1898, rev. ed. 1928; rpt. New York: Dover, 1962), pp. 288–289.

4. In 1980 my colleague Wilbur Hoppe and I built a compound pendulum as part of a workshop on mathematical models at the University of Wisconsin—Eau Claire. The patterns shown in figure 74 were produced with this device.

12

---

tan *x*

Of the numerous functions we encounter in elementary mathematics, perhaps the most remarkable is the tangent function. The basic facts are well known: $f(x) = \tan x$ has its zeros at $x = n\pi$ $(n = 0, \pm 1, \pm 2, \ldots)$, has infinite discontinuities at $x = (2n + 1)\pi/2$, and has period π (a function $f(x)$ is said to have a period P if P is the smallest number such that $f(x + P) = f(x)$ for all x in the domain of the function). This last fact is quite remarkable: the functions $\sin x$ and $\cos x$ have the common period 2π, yet their ratio, $\tan x$, reduces the period to π. When it comes to periodicity, the ordinary rules of the algebra of functions may not be valid: the fact that two functions f and g have a common period P does not imply that $f + g$ or fg too have the same period.[1]

As we saw in chapter 2, the tangent function has its origin in the "shadow reckoning" of antiquity. During the Renaissance it was resurrected—though without calling it "tangent"—in connection with the fledging art of perspective. It is a common experience that an object appears progressively smaller as it moves away from the observer. The effect is particularly noticeable when viewing a tall structure from the ground: as the angle of sight is elevated, features that are equally spaced vertically, such as the floors of a building, appear to be progressively shortened; and conversely, equal increments in the angle of elevation intercept the structure at points that are increasingly farther apart. A study by the famed Nürnberg artist Albrecht Dürer (1471–1528), one of the founders of perspective, clearly shows this effect (fig. 75).[2]

Dürer and his contemporaries were particularly intrigued by the extreme case of this phenomenon when the angle of elevation approaches 90° and the height seems to increase without limit. More intriguing still was the behavior of parallel lines in the plane: as they recede from the viewer they seem to get ever closer, ultimately converging on the horizon at a point called the

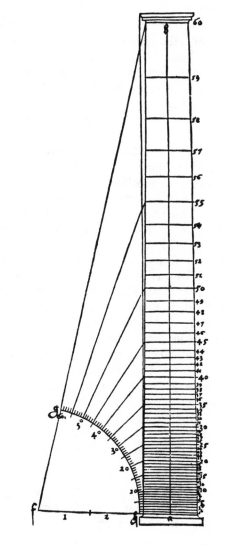

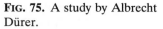

FIG. 75. A study by Albrecht Dürer.

"vanishing point." All these features can be traced to the behavior of $\tan x$ near $90°$. Today, of course, we say that $\tan x$ *tends* to infinity as x approaches $90°$, whereas *at* $90°$ it is undefined; but such subtleties were unknown to past generations, and until quite recently one could still find the statement "$\tan 90° = \infty$" in many trigonometry textbooks.

But let us return to mundane matters. Around 1580 Viète stated a beautiful theorem that, unfortunately, has all but disappeared from today's textbooks: the Law of Tangents. It says that in any triangle,

$$\frac{a+b}{a-b} = \frac{\tan(\alpha+\beta)/2}{\tan(\alpha-\beta)/2}. \tag{1}$$

This theorem follows from the Law of Sines ($a/\sin\alpha = b/\sin\beta = c/\sin\gamma$) and the identities $\sin\alpha \pm \sin\beta = 2\sin(\alpha \pm \beta)/2 \cdot \cos(\alpha \mp \beta)/2$, but in Viète's time it was regarded as an independent theorem.[3] It can be used to solve a triangle when two sides and the included angle are given (the *SAS* case). Normally one would use the Law of Cosines ($c^2 = a^2 + b^2 - 2ab\cos\gamma$) to find the missing side, and then find one or the other of the remaining angles using the Law of Sines. However, because the cosine law involves addition and subtraction, it does not lend itself easily to logarithmic computations—practically the only means of solving triangles (or most other computations) before the hand-held calculator became available. With the tangent law one could avoid this difficulty: since one angle, say γ, is given, one could find $(\alpha+\beta)/2$, and with the help of equation (1) and a table of tangents find $(\alpha-\beta)/2$; from these two results the angles α and β are found, and finally the two missing sides from the sine law. With a calculator, of course, this is no longer necessary, which may explain why the Law of Tangents has lost much of its appeal. Still, its elegant, symmetric form should be a good enough reason to resurrect it from oblivion, if not as a theorem then at least as an exercise. And for those who enjoy mathematical *misteakes*—correct results derived incorrectly (such as $16/64 = 1/4$)—the Law of Tangents provides ample opportunity: start with the right side of equation (1), "cancel" the 1/2 and the "tan" and replace Greek letters with corresponding Latin ones, and you get the left side.[4]

Other formulas involving $\tan x$ are just as elegant; for example, if α, β, and γ are the three angles of any triangle, we have

$$\tan\alpha + \tan\beta + \tan\gamma = \tan\alpha \cdot \tan\beta \cdot \tan\gamma. \tag{2}$$

This formula can be proved by writing $\gamma = 180° - (\alpha+\beta)$ and using the addition formula for the tangent. The formula is remarkable not only for its perfect symmetry, but also because it leads to an unexpected result. A well-known theorem from algebra says that if $x_1, x_2, \ldots, x_n$ are any positive numbers, then their arithmetic mean is never smaller than their geometric mean; that is,

$$\frac{x_1 + x_2 + \cdots + x_n}{n} \geq \sqrt[n]{x_1 x_2 \ldots x_n}.$$

Moreover, the two means are *equal* if, and only if, $x_1 = x_2 = \cdots = x_n$. Let us assume that our triangle is acute, so that all three angles have positive tangents. Then the theorem says that

$$\frac{\tan \alpha + \tan \beta + \tan \gamma}{3} \geq \sqrt[3]{\tan \alpha \cdot \tan \beta \cdot \tan \gamma}.$$

But in view of equation (2) this inequality becomes

$$\frac{\tan \alpha \cdot \tan \beta \cdot \tan \gamma}{3} \geq \sqrt[3]{\tan \alpha \cdot \tan \beta \cdot \tan \gamma}.$$

Cubing both sides gives

$$\tan \alpha \cdot \tan \beta \cdot \tan \gamma \geq \sqrt{27} = 3\sqrt{3}.$$

Thus in any acute triangle, the product (and sum) of the tangents of the three angles is never less than $3\sqrt{3} \approx 5.196$; and this minimum value is attained if, and only if, $\alpha = \beta = \gamma = 60°$, that is, when the triangle is equilateral.

If the triangle is obtuse, then one of the three angles has a negative tangent, in which case the theorem does not apply; however, since the obtuse angle can vary only from $90°$ to $180°$, its tangent ranges over the interval $(-\infty, 0)$, while the other two tangents remain positive and finite. Thus the product of the three tangents can assume any negative value.

Summing up, in any acute triangle we have $\tan \alpha \cdot \tan \beta \cdot \tan \gamma \geq 3\sqrt{3}$ (with equality if and only if the triangle is equilateral), and in any obtuse triangle we have $-\infty < \tan \alpha \cdot \tan \beta \cdot \tan \gamma < 0$.

✧ ✧ ✧

The tangent of a multiple of an angle provides us with another source of interesting formulas. In chapter 8 we obtained the formula

$$\tan (n+1)\alpha/2 = \frac{\sin \alpha + \sin 2\alpha + \cdots + \sin n\alpha}{\cos \alpha + \cos 2\alpha + \cdots + \cos n\alpha}.$$

We can make this slightly more useful by writing $\alpha/2 = \beta$ and $n + 1 = m$; then

$$\tan m\beta = \frac{\sin 2\beta + \sin 4\beta + \cdots + \sin 2(m-1)\beta}{\cos 2\beta + \cos 4\beta + \cdots + \cos 2(m-1)\beta}.$$

But even in this form the formula has only limited usefulness, because it expresses the tangent of a multiple of an angle in terms of the sines and cosines of other angles. It would be desirable if we could express $\tan n\alpha$ in terms of the *tangent* of α (and α alone, not its multiples). Fortunately, this can be done.

We start with the familiar addition formula for the tangent,

$$\tan(\alpha + \beta) = \frac{\tan\alpha + \tan\beta}{1 - \tan\alpha \cdot \tan\beta}.$$

From this we get

$$\tan 2\alpha = \tan(\alpha + \alpha) = \frac{2\tan\alpha}{1 - \tan^2\alpha},$$

$$\tan 3\alpha = \tan(2\alpha + \alpha) = \frac{\tan 2\alpha + \tan\alpha}{1 - \tan 2\alpha \cdot \tan\alpha}$$

$$= \frac{3\tan\alpha - \tan^3\alpha}{1 - 3\tan^2\alpha},$$

and so on. After a few steps a pattern begins to emerge: we discover that the coefficients are the same as those appearing in the expansion of $(1 + x)^n$ in powers of x—the familiar *binomial coefficients*—except that they alternate between the numerator and denominator in a zigzag pattern (beginning with the first term in the denominator) and their signs alternate in pairs.[5] Figure 76, taken from an early nineteenth-century trigonometry textbook, shows the pattern up to $\tan 7\alpha$. Taking the signs into account, we can arrange the coefficients in a "Pascal tangent triangle":

$$
\begin{array}{ccccccc}
 & & & 1 & & & \\
 & & 1 & & 1 & & \\
 & 1 & & 2 & & -1 & \\
1 & & 3 & & -3 & & -1 \\
\end{array}
$$

1	4		-6		-4		1
1	5	-10		-10		5		1

$$
\begin{array}{ccccccc}
1 & & 4 & & -6 & & -4 & & 1 \\
1 & & 5 & & -10 & & -10 & & 5 & & 1
\end{array}
$$

$$\cdots$$

(the top entry is 1 because $\tan 0 = 0 = 0/1$). We notice that all entries in the first two diagonal lines (counting from left to right) are positive, those of the next two lines are negative, and so on.

The presence of the binomial coefficients in a formula that appears to have nothing to do with the expansion of $(1 + x)^n$ is

CHAPTER XI.

$$\tan a = \tan a$$

$$\tan 2a = \frac{2 \tan a}{1 - \tan^2 a}$$

$$\tan 3a = \frac{3 \tan a - \tan^3 a}{1 - 3 \tan^2 a}$$

$$\tan 4a = \frac{4 \tan a - 4 \tan^3 a}{1 - 6 \tan^2 a + \tan^4 a}$$

$$\tan 5a = \frac{5 \tan a - 10 \tan^3 a + \tan^5 a}{1 - 10 \tan^2 a + 5 \tan^4 a}$$

$$\tan 6a = \frac{6 \tan a - 20 \tan^3 a + 6 \tan^5 a}{1 - 15 \tan^2 a + 15 \tan^4 a - \tan^6 a}$$

$$\tan 7a = \frac{7 \tan a - 35 \tan^3 a + 21 \tan^5 a - \tan^7 a}{1 - 21 \tan^2 a + 35 \tan^4 a - 7 \tan^6 a}$$

$$\tan 8a = \text{\&c.}$$

§ 45. If we consider the foregoing formulæ for the sine and cosine of the multiple angles expressed wholly in terms of the sines and cosines of the simple angles, and their successive powers, both in relation to the order in which these powers, and to that in which their coefficients, occur, we shall perceive, that: for every corresponding multiple of the sine and cosine, beginning at the first term of the cosine, thence passing to the first term of the sine, then from the second term of the cosine to the second of the sine, and so on to the end; we have all the terms of the binomial in regular order, as well for the powers of cosine *a*, and sine *a*, as for their numeric coefficients; with this difference only, that a regular change of the signs, +, and —, takes place separately, in each of the series.

The same law holds good in the case of the tangents, as far as regards the coefficients; and the powers of the tangents follow in a regular order, from the numerator to the denominator, alternately.

FIG. 76. Expansion of $\tan na$ in powers of $\tan a$. From an early nineteenth-century trigonometry book.

a result of De Moivre's theorem (see p. 83),

$$(\cos \alpha + i \sin \alpha)^n = \cos n\alpha + i \sin n\alpha, \tag{3}$$

where $i = \sqrt{-1}$. If we expand the left side of equation (3) according to the binomial theorem and equate the real and imaginary parts with those on the right, we get expressions for $\cos n\alpha$ and $\sin n\alpha$ in terms of $\cos^{n-k} \alpha \cdot \sin^k \alpha$, where $k = 0, 1, 2, \ldots, n$; from these the formula for $\tan n\alpha$ easily follows:

$$\tan n\alpha = \frac{n \tan \alpha - {}^nC_3 \tan^3 \alpha + {}^nC_5 \tan^5 \alpha - + \cdots}{1 - {}^nC_2 \tan^2 \alpha + {}^nC_4 \tan^4 \alpha - + \cdots}, \tag{4}$$

where the symbol ${}^nC_k$—also denoted by $\binom{n}{k}$—stands for

$${}^nC_k = \frac{n \cdot (n-1) \cdot (n-2) \cdot \ldots \cdot (n-k+1)}{k!}. \tag{5}$$

For example, ${}^4C_3 = (4 \cdot 3 \cdot 2)/(1 \cdot 2 \cdot 3) = 4$. Note that the expression on the right side of equation (5) is also equal to $n!/[k! \cdot (n-k)!]$, so we have ${}^nC_k = {}^nC_{n-k}$ (in the example just given, ${}^4C_1 = 4!/(3! \cdot 1!) = 4 = {}^4C_3$). Because of this, the binomial coefficients are symmetric—they are the same whether one expands $(1+x)^n$ in ascending or descending powers of x.

✧ ✧ ✧

Because the expansion of $(1+x)^n$ for positive integral n involves $(n+1)$ terms, the series appearing in the numerator and denominator of equation (4) are finite sums. But all six trigonometric functions can also be represented by *infinite* expressions, specifically power series and infinite products. The power series for $\sin x$ and $\cos x$ are

$$\sin x = x - x^3/3! + x^5/5! - + \cdots$$

and

$$\cos x = 1 - x^2/2! + x^4/4! - + \cdots.$$

These series were already known to Newton, but it was the great Swiss mathematician Leonhard Euler (1707–1783) who used them to derive a wealth of new results. Euler regarded power series as "infinite polynomials" that obey the same rules of algebra as do ordinary, finite polynomials. Thus, he argued, just as a polynomial of degree n can be written as a product of n (not necessarily different) factors of the form $(1 - x/x_i)$, where x_i are the roots, or zeros, of the polynomial,[6] so can the function $\sin x$ be written as an *infinite product*

$$\sin x = x(1 - x^2/\pi^2)(1 - x^2/4\pi^2)(1 - x^2/9\pi^2) \cdots. \tag{6}$$

Here each quadratic factor $(1 - x^2/n^2\pi^2)$ is the product of the two linear factors $(1 - x/n\pi)$ and $(1 + x/n\pi)$ resulting from the zeros of $\sin x$, $x_n = \pm n\pi$, whereas the single factor x results from the zero $x = 0$.[7] One surprising consequence of equation (6) results if we substitute $x = \pi/2$:

$$1 = (\pi/2) \cdot (1 - 1/4) \cdot (1 - 1/16) \cdot (1 - 1/36) \cdot \cdots .$$

Simplifying each term and solving for $\pi/2$, we get the infinite product

$$\frac{\pi}{2} = \frac{2}{1} \cdot \frac{2}{3} \cdot \frac{4}{3} \cdot \frac{4}{5} \cdot \frac{6}{5} \cdot \frac{6}{7} \cdots . \tag{7}$$

This famous formula is named after John Wallis (see p. 51), who discovered it in 1655 through a daring interpolation process.[8]

The infinite product for $\cos x$ is

$$\cos x = (1 - 4x^2/\pi^2)(1 - 4x^2/9\pi^2)(1 - 4x^2/25\pi^2) \cdots , \tag{8}$$

where $x_i = \pm\pi/2, \pm 3\pi/2, \ldots$ are the zeros of $\cos x$ (here again each quadratic factor is the product of two linear factors). If we now divide equations (6) by (8), we get an analytic expression for $\tan x$:

$$\tan x = \frac{x(1 - x^2/\pi^2)(1 - x^2/4\pi^2)(1 - x^2/9\pi^2) \cdots}{(1 - 4x^2/\pi^2)(1 - 4x^2/9\pi^2)(1 - 4x^2/25\pi^2) \cdots}. \tag{9}$$

This expression, however, is rather cumbersome. To simplify it, let us use a technique familiar from the integral calculus—the decomposition of a rational function into *partial fractions*. We express the right side of equation (9) as an infinite sum of fractions, each with a denominator equal to one *linear* factor in the denominator of equation (9):

$$\tan x = \frac{A_1}{1 - 2x/\pi} + \frac{B_1}{1 + 2x/\pi} + \frac{A_2}{1 - 2x/3\pi}$$

$$+ \frac{B_2}{1 + 2x/3\pi} + \cdots . \tag{10}$$

To find the coefficients of this decomposition, let us "clear fractions": we multiply both sides of equation (10) by the product of all denominators (that is, by $\cos x$), and equate the result to the numerator of equation (9) (that is, to $\sin x$):

$$x(1 - x/\pi)(1 + x/\pi)(1 - x/2\pi)(1 + x/2\pi) \cdots$$

$$= A_1(1 + 2x/\pi)(1 - 2x/3\pi)(1 + 2x/3\pi) \cdots$$

$$+ B_1(1 - 2x/\pi)(1 - 2x/3\pi)(1 + 2x/3\pi) \cdots$$

$$+ \cdots . \tag{11}$$

Note that in each term on the right side of equation (11) exactly one factor is missing—the denominator of the corresponding coefficient in equation (10) (just as it is when finding an ordinary common denominator).

Now equation (11) is an identity in x—it holds true for any value of x we choose to put in it. To find A_1, let us choose $x = \pi/2$; this will "annihilate" all terms except the first, giving us

$$(\pi/2) \cdot (1/2) \cdot (3/2) \cdot (3/4) \cdot (5/4) \cdot \cdots$$
$$= A_1 \cdot 2 \cdot (2/3) \cdot (4/3) \cdot (4/5) \cdot (6/5) \cdot \cdots.$$

Solving for A_1, we get

$$A_1 = (\pi/2) \cdot (1/2)^2 \cdot (3/2)^2 \cdot (3/4)^2 \cdot (5/4)^2 \cdot \cdots$$
$$= (\pi/2) \cdot [(1/2) \cdot (3/2) \cdot (3/4) \cdot (5/4) \cdot \cdots]^2.$$

But the expression inside the brackets is exactly the reciprocal of Wallis's product, that is, $2/\pi$; we thus have

$$A_1 = (\pi/2) \cdot (2/\pi)^2 = 2/\pi.$$

To find B_1 we follow the same process except that now we put $x = -\pi/2$ in equation (11); this gives us $B_1 = -2/\pi = -A_1$. The other coefficients are obtained in a similar way;[9] we find that $A_2 = 2/(3\pi) = -B_2$, $A_3 = 2/(5\pi) = -B_3$, and in general

$$A_i = \frac{2}{(2i - 1)\pi} = -B_i.$$

Putting these coefficients back into equation (10) and combining the terms in pairs, we get our grand prize, the decomposition of $\tan x$ into partial fractions:

$$\tan x = 8x \left[\frac{1}{\pi^2 - 4x^2} + \frac{1}{9\pi^2 - 4x^2} \right.$$
$$\left. + \frac{1}{25\pi^2 - 4x^2} + \cdots \right]. \tag{12}$$

This remarkable formula shows directly that $\tan x$ is undefined at $x = \pm\pi/2, \pm3\pi/2, \ldots$; these, of course, are precisely the vertical asymptotes of $\tan x$.

✧   ✧   ✧

Now that we have spent so much labor on establishing equation (12), let us reap some benefits from it. Since the equation holds for all x except $(2n+1)\pi/2$, $n = 0, \pm 1, \pm 2, \ldots$, let us put in it some special values. We start with $x = \pi/4$:

$$\tan \pi/4 = 1 = 8(\pi/4)[1/(\pi^2 - \pi^2/4) + 1/(9\pi^2 - \pi^2/4)$$
$$+ 1/(25\pi^2 - \pi^2/4) + \cdots]$$
$$= (8/\pi)[1/3 + 1/35 + 1/99 + \cdots].$$

Each term inside the brackets is of the form $1/[4(2n-1)^2 - 1] = 1/(4n-3)(4n-1) = (1/2)[1/(4n-3) - 1/(4n-1)]$, $n = 1, 2, 3, \ldots$; we thus have

$$1 = (4/\pi)[1 - 1/3 + 1/5 - 1/7 + - \cdots],$$

from which we get

$$\frac{\pi}{4} = 1 - \frac{1}{3} + \frac{1}{5} - \frac{1}{7} + - \cdots. \tag{13}$$

This famous formula was discovered in 1671 by the Scottish mathematician James Gregory (1638–1675), who derived it from the power series for the inverse tangent, $\tan^{-1} x = x - x^3/3 + x^5/5 - + \cdots$, from which equation (13) follows by substituting $x = 1$. Leibniz discovered the same formula independently in 1674, and it is often named after him.[10] It was one of the first results of the newly invented differential and integral calculus, and it caused Leibniz much joy.

The remarkable thing about the Gregory-Leibniz series—as also Wallis's product—is the unexpected connection between π and the integers. However, because of its very slow rate of convergence, this series is of little use from a computational point of view: it requires 628 terms to approximate π to just two decimal places—an accuracy far worse than that obtained by Archimedes, using the method of exhaustion, two thousand years earlier. Nevertheless, the Gregory-Leibniz series marks a milestone in the history of mathematics as the first of numerous infinite series involving π to be discovered in the coming years.

Next, let us use equation (12) with $x = \pi$:

$$\tan \pi = 0 = 8\pi[1/(-3\pi^2) + 1/(5\pi^2) + 1/(21\pi^2) + \cdots].$$

Canceling out $8/\pi$ and moving the negative term to the left side of the equation, we get

$$1/5 + 1/21 + 1/45 + \cdots = 1/3.$$

Perhaps somewhat disappointingly, we arrived at a series that does not involve π.[11] But excitement returns when we try to put

$x = 0$ in equation (12). At first we merely get the indeterminate equation $0 = 0$, but we can go around this difficulty by dividing both sides of the equation by x and then letting x *approach zero*. On the left side we get

$$\lim_{x \to 0} \frac{\tan x}{x} = \left(\lim_{x \to 0} \frac{\sin x}{x} \right) \cdot \left(\lim_{x \to 0} \frac{1}{\cos x} \right) = 1 \cdot 1 = 1.$$

Thus equation (12) becomes

$$1 = (8/\pi^2)\left[\frac{1}{1} + \frac{1}{9} + \frac{1}{25} + \cdots \right]$$

or

$$\frac{\pi^2}{8} = \frac{1}{1^2} + \frac{1}{3^2} + \frac{1}{5^2} + \cdots. \tag{14}$$

This last formula is as remarkable as the Gregory-Leibniz series, but we can derive from it an even more interesting result. We will again do this in a nonrigorous way, in the spirit of Euler's daring forays into the world of infinite series (a more rigorous proof will be given in chapter 15). Our task is to find the sum of the reciprocals of the squares of *all* positive integers, even and odd; let us denote this sum by S:[12]

$$S = \frac{1}{1^2} + \frac{1}{2^2} + \frac{1}{3^2} + \frac{1}{4^2} + \cdots$$

$$= \left(\frac{1}{1^2} + \frac{1}{3^2} + \cdots \right) + \left(\frac{1}{2^2} + \frac{1}{4^2} + \cdots \right)$$

$$= \left(\frac{1}{1^2} + \frac{1}{3^2} + \cdots \right) + \frac{1}{4}\left(\frac{1}{1^2} + \frac{1}{2^2} + \cdots \right)$$

$$= \frac{\pi^2}{8} + \frac{1}{4}S.$$

From this we get $(3/4)S = \pi^2/8$, and finally

$$S = \frac{1}{1^2} + \frac{1}{2^2} + \frac{1}{3^2} + \cdots = \frac{\pi^2}{6}. \tag{15}$$

Equation (15) is one of the most celebrated formulas in all of mathematics; it was discovered by Euler in 1734 in a flash of ingenuity that would defy every modern standard of rigor. Its discovery solved one of the great mysteries of the eighteenth century: it had been known for some time that the series converges, but the value of its sum eluded the greatest mathematicians of the time, among them the Bernoulli brothers.[13]

✧ ✧ ✧

We consider one more infinite series discovered by Euler. We begin with the double-angle formula for the cotangent,

$$\cot 2x = \frac{1 - \tan^2 x}{2 \tan x} = \frac{\cot x - \tan x}{2}.$$

Starting with an arbitrary angle $x \neq n\pi/2$ and applying the formula repeatedly, we get

$$\cot x = \frac{1}{2}(\cot x/2 - \tan x/2)$$

$$= \frac{1}{4}(\cot x/4 - \tan x/4) - \frac{1}{2}\tan x/2$$

$$= \frac{1}{8}(\cot x/8 - \tan x/8) - \frac{1}{4}\tan x/4 - \frac{1}{2}\tan x/2$$

$$= \cdots$$

$$= \frac{1}{2^n}(\cot x/2^n - \tan x/2^n)$$

$$- \frac{1}{2^{n-1}}\tan x/2^{n-1} - \cdots - \frac{1}{2}\tan x/2.$$

As $n \to \infty$, $[\cot(x/2^n)]/2^n$ tends to $1/x$,[14] so we get

$$\cot x = \frac{1}{x} - \sum_{n=1}^{\infty} \frac{1}{2^n}\tan\frac{x}{2^n}$$

or

$$\frac{1}{x} - \cot x = \frac{1}{2}\tan\frac{x}{2} + \frac{1}{4}\tan\frac{x}{4} + \frac{1}{8}\tan\frac{x}{8} + \cdots. \tag{16}$$

This little-known formula is one more of hundreds of formulas involving infinite processes to emerge from Euler's imaginative mind. And behind it a surprise is hiding: if we put in it $x = \pi/4$, we get

$$4/\pi - 1 = (1/2)\tan \pi/8 + (1/4)\tan \pi/16 + \cdots.$$

Replacing the 1 on the left side with $\tan \pi/4$, moving all the tangent terms to the right side, and dividing the equation by 4, we get

$$\frac{1}{\pi} = \frac{1}{4}\tan\frac{\pi}{4} + \frac{1}{8}\tan\frac{\pi}{8} + \frac{1}{16}\tan\frac{\pi}{16} + \cdots. \tag{17}$$

Equation (17) must surely rank among the most beautiful in mathematics, yet it hardly ever shows up in textbooks. Moreover, the series on the right side converges extremely rapidly (note that the coefficients *and* the angles decrease by a factor of 1/2 with each term), so we can use equation (17) as an efficient means to approximate π: it takes just twelve terms to obtain π to six decimal places, that is, to one-millionth; four more terms will increase the accuracy to one billionth.[15]

We have followed Euler's spirit in handling equations such as (6) and (9) as if they were finite expressions, subject to the rules of ordinary algebra. Euler lived in an era of carefree mathematical exploration when formal manipulation of infinite series was a normal practice; the questions of convergence and limit were not yet fully understood and were thus by and large ignored. Today we know that these questions are crucial to all infinite processes, and that ignoring them can lead to false results.[16] To quote George F. Simmons in his excellent calculus textbook, "These daring speculations are characteristic of Euler's unique genius, but we hope that no student will suppose that they carry the force of rigorous proof."[17]

NOTES AND SOURCES

1. A simple example of this is given by the functions $f(x) = \sin x$, $g(x) = 1 - \sin x$. Each has period 2π, yet their sum, $f(x) + g(x) = 1$, being a constant, has any real number as period.

2. On Dürer's mathematical work, see Julian Lowell Coolidge, *The Mathematics of Great Amateurs* (1949; rpt. New York: Dover, 1963), chap. 5, and Dan Pedoe, *Geometry and the Liberal Arts* (New York: St. Martin's Press, 1976), chap. 2.

3. An equivalent form of the law, in which the left side of the equation is replaced by $(\sin \alpha + \sin \beta)/(\sin \alpha - \sin \beta)$, was already known to Regiomontanus around 1464, but curiously he did not include it—nor any other applications of the tangent function—in his major treatise, *On Triangles* (see p. 44). As for other discoverers of the Law of Tangents, see David Eugene Smith, *History of Mathematics* (1925; rpt. New York: Dover, 1958), vol. 2, pp. 611 and 631.

4. Trigonometry abounds in such examples. We have seen one in chapter 8 in connection with the summation formula for $\sin \alpha + \sin 2\alpha + \cdots + \sin n\alpha$. Another example is the identity $\sin^2 \alpha - \sin^2 \beta = \sin (\alpha + \beta) \cdot \sin (\alpha - \beta)$, which can be "proved" by writing the left side as $\sin (\alpha^2 - \beta^2) = \sin [(\alpha + \beta) \cdot (\alpha - \beta)] = \sin (\alpha + \beta) \cdot \sin (\alpha - \beta)$.

5. We can take the signs into account by considering the expansion of $(1 + ix)^n$, where $i = \sqrt{-1}$.

6. This is equivalent to the more familiar factorization into factors of the form $(x - x_i)$. For example, the zeros of the polynomial $f(x) =$

$x^2 - x - 6$ are -2 and 3, so we have $f(x) = (x + 2)(x - 3) = -6(1 + x/2)(1 - x/3)$. In general, a polynomial $f(x) = a_n x^n + a_{n-1} x^{n-1} + \cdots + a_1 x + a_0$ can be written either as $a_n(x - x_1) \cdot \ldots \cdot (x - x_n)$, where a_n is the leading coefficient or as $a_0(1 - x/x_1) \cdot \ldots \cdot (1 - x/x_n)$, where a_0 is the constant term.

7. Actually Euler discarded the root $x = 0$ and therefore obtained the infinite product for $(\sin x)/x$. See Morris Kline, *Mathematical Thought from Ancient to Modern Times* (New York: Oxford University Press, 1972), vol. 2, pp. 448–449.

8. See *A Source Book in Mathematics*, 1200–1800, ed. D. J. Struik (Cambridge, Mass.: Harvard University Press, 1969), pp. 244–253. For a rigorous proof of equations (6) and (7), see Richard Courant, *Differential and Integral Calculus* (London: Blackie & Son, 1956), vol. 1, pp. 444–445 and 223–224. Other infinite products for π can also be derived from equation (6); for example, by putting $x = \pi/6$ we get

$$\frac{\pi}{3} = \frac{6}{5} \cdot \frac{6}{7} \cdot \frac{12}{11} \cdot \frac{12}{13} \cdot \frac{18}{17} \cdot \frac{18}{19} \cdots$$

$$= \prod_{n=1}^{\infty} (6n)^2 / [(6n)^2 - 1],$$

which actually converges faster than Wallis's product (it takes 55 terms of this product to approximate π to two decimal places, compared to 493 terms of Wallis's product).

9. However, the resulting numerical products become more complicated as i increases. Fortunately there is an easier way to find the coefficients: the left side of equation (11) is $\sin x$, while each term on the right side is equal to $\cos x$ divided by the missing denominator of that term. Thus, to find A_2 we put $x = 3\pi/2$ in equation (11); this will "annihilate" all terms except that of A_2, and for the surviving term we have

$$(\sin x)_{x=3\pi/2} = A_2 \left[\frac{\cos x}{1 - 2x/3\pi} \right]_{x=3\pi/2}.$$

The left side equals -1, but on the right side we get the indeterminate expression $0/0$. To evaluate it, we use L'Hospital's rule and transform it into $A_2[(-\sin x)/(-2/3\pi)]_{x=3\pi/2} = -(3\pi/2)A_2$. We thus get $A_2 = 2/(3\pi)$. The other coefficients can be found in the same way.

10. See Petr Beckmann, *A History of π* (Boulder, Colo.: Golem Press, 1977), pp. 132–133; for Leibniz's proof, see George F. Simmons, *Calculus with Analytic Geometry* (New York: McGraw-Hill, 1985), pp. 720–721. The series for $\tan^{-1} x$ can be obtained by writing the expression $1/(1 + x^2)$ as a power series $1 - x^2 + x^4 - + \cdots$ (a geometric series with the quotient $-x^2$) and integrating term by term.

11. That the series $1/5 + 1/21 + 1/45 + \cdots$ converges to $1/3$ can be confirmed by noting that each term has the form $1/[(2n + 1)^2 - 4] = 1/(2n - 1)(2n + 3) = (1/4)[1/(2n - 1) - 1/(2n + 3)]$; thus the series becomes

$$\tfrac{1}{4}[(1 - \tfrac{1}{5}) + (\tfrac{1}{3} - \tfrac{1}{7}) + (\tfrac{1}{5} - \tfrac{1}{9}) + (\tfrac{1}{7} - \tfrac{1}{11}) + \cdots].$$

This is a "telescopic" series in which all terms except the first and third cancel out, resulting in the sum $(1/4)(1 + 1/3) = 1/3$.

12. Assuming, of course, that the series converges. It is proved in calculus texts that the series $\sum_{n=1}^{\infty} 1/n^k$, where k is a real number, converges for all $k > 1$, and diverges for $k \leq 1$. In our case $k = 2$, hence S converges.

13. See Simmons, *Calculus*, pp. 722–723 (Euler's proof) and pp. 723–725 (a rigorous proof). One would expect that the series $\sum_{n=1}^{\infty} 1/n^2$ converges much faster than the Gregory-Leibniz series because all its terms are positive and involve the squares of the integers. Surprisingly, this is not so: it takes 600 terms to approximate π to two decimal places, compared to 628 terms of the Gregory-Leibniz series.

14. $\lim_{n\to\infty}[\cot(x/2^n)]/2^n = (1/x)\lim_{n\to\infty}[(x/2^n)\cot(x/2^n)] = 1/x$, the last result following from $\lim_{t\to\infty}(1/t)\cot(1/t) = \lim_{u\to0} u \cot u = \lim_{u\to0} u/\tan u = 1$, where $u = 1/t$.

15. One may raise the objection that equation (17) expresses π in terms of itself, since the angles in the tangent terms are in radians. However, the trigonometric functions are "immune" to the choice of the angular unit; using degrees instead of radians, equation (17) becomes $1/\pi = (1/4)\tan 45° + (1/8)\tan 45°/2 + \cdots$.

16. On this subject see Kline, *Mathematical Thought*, vol. 2, pp. 442–454 and 460–467. See also my book, *To Infinity and Beyond: A Cultural History of the Infinite* (Princeton, N.J.: Princeton University Press, 1991), pp. 32–33 and 36–39.

17. Simmons, *Calculus*, p. 723.

13

A Mapmaker's Paradise

What's the good of Mercator's North Poles and
Equators, Tropics, Zones and Meridian Lines?
So the Bellman would cry: and the crew would reply,
"They are merely conventional signs!"
—Lewis Carroll (Charles Dodgson), *The Hunting of*
the Snark (1876)

From the sublime beauty of Euler's formulas we now turn to a more mundane matter: the science of map making. It is common knowledge that one cannot press the peels of an orange against a table without tearing them apart: no matter how carefully one tries to do the job, some distortion is inevitable. Surprisingly, it was not until the middle of the eighteenth century that this fact was proved mathematically, and by none other than Euler: his theorem says that it is impossible to map a sphere onto a flat sheet of paper without distortion. Had the earth been a cylinder, or a cone, the mapmaker's task would have been easier: these surfaces are *developable*—they can be flattened without shrinking or stretching. This is because these surfaces, though curved, have essentially the geometry of a plane. But the underlying geometry of a sphere is fundamentally different from that of a plane; consequently, one cannot create a map of the earth that faithfully reproduces *all* its features.

To cope with this problem, cartographers have devised a variety of *map projections*—functions (in the mathematical sense) that assign to every point on the sphere an "image" point on the map. The choice of a particular projection depends on the intended purpose of the map; one map may show the correct distance between two points on the globe (of course up to a scaling factor), another the relative area of countries, and yet another the direction between two points. But preserving any of these features always comes at the expense of other features: every map projection is a compromise between conflicting demands.

The simplest of all is the *cylindrical projection*: imagine the earth—represented by a perfectly spherical globe of radius R— to be wrapped in a cylinder touching it at the equator (fig. 77). Imagine further that rays of light emanate from the center of the globe in all directions. A point P on the globe is then projected onto a point P', the "shadow" or "image" of P on the cylinder. When the cylinder is unwrapped, we get a flat map of the entire earth—or *almost* the entire earth: the North and South Poles, being on the axis of the cylinder, have their images at infinity.

Clearly the cylindrical projection maps all circles of longitude (meridians) onto equally spaced vertical lines, while circles of latitude (or "parallels," as they are known in geography) show as horizontal lines whose spacing increases with latitude. In order to find the relation between a point P and its image P', we must first express the location of P in terms of its *longitude* (measured eastward or westward along the equator from the prime meridian through Greenwich, England) and *latitude* (measured northward or southward from the equator along any meridian). Denoting the longitude and latitude of P by the Greek letters λ (lambda) and ϕ and the coordinates of P' by x and y, we have

$$x = R\lambda, \ y = R \tan \phi. \tag{1}$$

The most striking feature of the cylindrical projection is the excessive north-south stretching at high latitudes, resulting in a drastic distortion of the shape of continents (fig. 78); this, of course, is a consequence of the presence of $\tan \phi$ in the second of equations (1). The cylindrical projection has often been

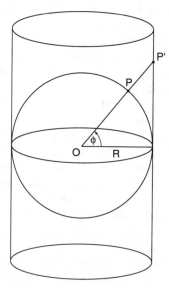

FIG. 77. Cylindrical projection of the globe.

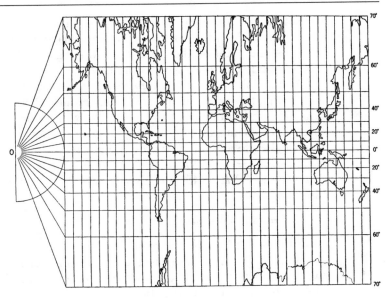

FIG. 78. World map on a cylindrical projection.

confused with Mercator's projection, which it resembles super-
ficially; however, except for the fact that both use a rectangular
grid, the two projections are based on entirely different principles, as we will shortly see.

A second projection, known already to Hipparchus in the second century B.C., is the *stereographic projection*. We place the
globe on a flat sheet of paper, touching it at the South Pole
S (fig. 79). We now connect every point *P* on the globe by a
straight line to the *North* Pole *N* and extend this line until it
meets the plane of the map at the point *P'*; *P'* is the image of *P*
under the projection.

The stereographic projection shows all meridians as straight
lines radiating from the South Pole *S*, while circles of latitude
show as concentric circles around *S*. The equator goes over to
a circle *e*, which we may think of as the unit circle. The entire
northern hemisphere is then mapped onto the exterior of *e* and

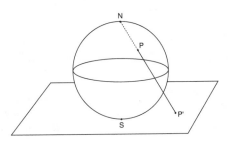

FIG. 79. Stereographic
projection from the North
Pole.

the southern hemisphere onto its interior. The closer a point is to the North Pole, the farther out will its image be on the map. There is one point on the globe with no image on the map: the North Pole itself. Its image is at infinity.

Let the globe have a unit *diameter*; this will ensure that circle e (the equator on the map) has a unit radius. Consider now a point P with latitude ϕ on the globe. We wish to determine the location of its image P' on the map. Figure 80 shows a cross section of the globe, with E representing a point on the equator; we have $SN = 1$, $\angle ONE = 45°$, $\angle EOP = \phi$, and $\angle ENP = \phi/2$. Therefore, $\angle ONP = (45° + \phi/2)$ and thus P' is located at a distance

$$SP' = \tan(45° + \phi/2) \tag{2}$$

from the South Pole on the map.

Equation (2) leads to an interesting result. Let P and Q be two points with the same longitude but opposite latitudes on the globe. How will their images be related on the map? Replacing ϕ with $-\phi$ in equation (2), we have

$$SQ' = \tan(45° - \phi/2) = \frac{1 - \tan\phi/2}{1 + \tan\phi/2}$$
$$= \frac{1}{\tan(45° + \phi/2)} = \frac{1}{SP'}.$$

Thus $SP' \cdot SQ' = 1$. Two points in the plane fulfilling this condition are said to be *inverse points* with respect to the unit circle; thus the stereographic projection sends two points with equal but opposite latitudes on the globe to two mutually inverse points on the map. This allows us to deduce all properties of the stereographic projection from the theory of inversion. It is known, for example, that the angle of intersection of two curves remains unchanged, or *invariant*, when each of the curves is subjected to inversion. From this one can show that the stereographic projection is direction-preserving, or *conformal*; that is, small regions on the globe preserve their shape on the map (hence the name conformal).[1] Figure 81 shows the northern

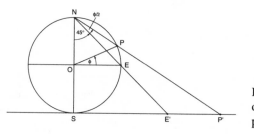

FIG. 80. Geometry of the stereographic projection.

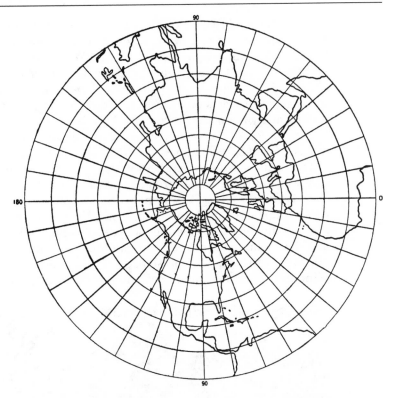

FIG. 81. The northern hemisphere on a stereographic projection.

hemisphere on a stereographic projection in which the globe touches the map at the North (rather than South) Pole; clearly the shape of continents is close to what it is on the globe.

✦   ✦   ✦

In chapter 10 we encountered the azimuthal equidistant map, which shows the true distance and direction from a given fixed point to any other point on the globe. However, distances and directions between any other two points are *not* preserved on this map; hence its usefulness for navigation is greatly limited. One would rather have a map that shows the correct direction, or compass bearing, from *any* point on the globe to *any* other point. But up until the middle of the sixteenth century no such map existed.

Imagine you are the navigator of a boat about to leave port headed in a certain direction. You set your compass at your chosen bearing, say 45°east of north, and then follow that bearing steadfastly, ignoring—for the sake of argument—any land masses that might be in your way. What path will you follow? For many years it had been believed that a path of constant bearing—known as a *rhumb line* or *loxodrome*[2]—is an arc of a great circle (see p. 136). But the Portuguese Pedro Nuñes (or Nonius, 1502–1578) showed that the rhumb line is actually a spiral curve that gets ever closer to either pole, winding around it indefinitely but never reaching it. The Dutch artist Maurits C. Escher (1898–1972) has depicted the rhumb line in one of his works, *Sphere Surface with Fish* (1958), shown in figure 82.

The challenge that faced cartographers in the sixteenth century was to design a map projection *that would show all rhumb lines as straight lines*. Such a map would enable a navigator to join his points of departure and destination by a straight line, measure the angle, or bearing, between this line and the north, and then follow this bearing at sea. On all existing projections, however, a straight-line course laid off on the map did not correspond to a rhumb line at sea. As a result, navigation was an

FIG. 82. Maurits C. Escher's *Sphere Surface with Fish* (1958). ©1997 Cordon-Art-Baarn-Holland. All Rights reserved.

extremely tricky business—and a risky one too, for many lives were lost due to ships failing to reach their destination. It befell a Flemish mapmaker to come to the mariners' rescue.

✧   ✧   ✧

Gerardus Mercator, by general consensus the most famous mapmaker in history, was born Gerhard Kremer in Rupelmonde in Flanders (now Belgium but then part of Holland) on March 5, 1512. Only twenty years earlier Christopher Columbus had made his historic voyage to the New World, and young Kremer's imagination was fired by the new geographical discoveries. He entered the University of Louvain in 1530, and soon after graduating established himself as one of Europe's leading mapmakers and instrument designers. As was customary for learned men at the time, he Latinized his name to Mercator ("merchant," a literal translation of the Dutch word *kramer*), and it is by this name that he has been known ever since.

Mercator's promising career was threatened in 1544 when he was arrested as a heretic for practicing Protestantism in a Catholic country. He barely saved his life and subsequently fled to neighboring Duisburg (now Germany), where he settled in 1552. He remained there for the rest of his life.[3]

Before Mercator, mapmakers decorated their charts with fanciful mythological figures and imaginary lands of their own creation: their maps were more works of art than true representations of the earth. Mercator was the first to base his maps entirely on the most recent data collected by explorers, and in so doing he transformed cartography from an art to a science. He was also one of the first to bound in one volume a collection of separate maps, calling it an "atlas" in honor of the legendary globe-holding mythological figure that decorated the title page; this work was published in three parts, the last appearing in 1595, one year after his death.[4]

It was in 1568 that Mercator set himself the task of inventing a new map projection that would answer the mariner's needs and change global navigation from a haphazard, risky endeavor to an exact science. From the outset he was guided by two principles: the map was to be laid out on a rectangular grid, with all circles of latitude represented by horizontal lines parallel to the equator and equal to it in length, and all meridians showing as vertical lines perpendicular to the equator; and the map would be *conformal*, for only such a map could preserve the true direction between any two points on the globe.

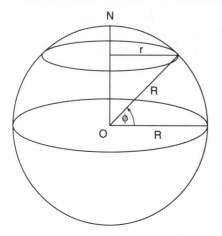

FIG. 83. Circle of latitude ϕ on the globe.

Now on the globe, the cirlces of latitude decrease in size as their latitude increases, until they shrink to a point at either pole. But on Mercator's map these same circles show as horizontal lines of *equal* length. Consequently, each parallel on the map is stretched horizontally (i.e., in an east-west direction) by a factor that depends on the latitude of that parallel. Figure 83 shows a circle of latitude ϕ. Its circumference is $2\pi r = 2\pi R \cos \phi$ on the globe, whereas on the map its length is $2\pi R$; it is thus stretched by a factor $2\pi R/(2\pi R \cos \phi) = \sec \phi$. Note that this stretching factor is *a function of* ϕ: the higher the latitude, the greater the stretching ratio, as table 6 shows.

And now Mercator was ready to produce his trump card: in order for the map to be conformal, the east-west stretching of the parallels must be accompanied by an equal north-south stretching of the spacing *between* the parallels, and this north-south stretching progressively increases as one goes to higher latitudes. In other words, the degrees of latitude, which on the globe are equally spaced along each meridian, must gradually be increased on the map (fig. 84). This is the key principle behind his map.

However, in order to implement this plan, the spacing between successive parallels had first to be determined. Exactly how Mercator did this is not known (and is still being debated

TABLE 6.   Sec ϕ for Some Selected Latitudes

ϕ	0°	15°	30°	45°	60°	75°	80°	85°	87°	89°	90°
sec ϕ	1.00	1.04	1.15	1.41	2.00	3.86	5.76	11.47	19.11	57.30	∞

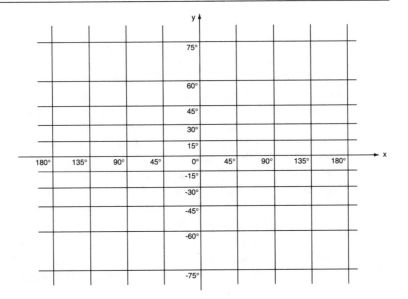

FIG. 84. The Mercator grid.

by historians of cartography);[5] he left no written record of his method except for the following brief explanation which was printed on his map:

In making this representation of the world, we had to spread on a plane the surface of the sphere in such a way that the positions of places shall correspond on all sides with each other both in true direction and in distance. . . .[6] With this intention we had to employ a new proportion and a new arrangement of the meridians with reference to the parallels. . . . For these reasons we have progressively increased the degrees of latitude towards each pole in proportion to the lengthening of the parallels with reference to the equator.[7]

Even this vague explanation makes it clear that Mercator had a full grasp of the mathematical principles underlying his map. Having created his grid, it now remained for him to put skin to skeleton—to superimpose on this grid the outline of the continents as they were known in his time. He published his world map (or "chart," as mariners prefer to call it) in 1569 under the title *New and Improved Description of the Lands of the World, amended and intended for the Use of Navigators*. It was a huge map, printed on twenty-one sections and measuring 54 by 83 inches. It is one of the most treasured cartographical artifacts of all time: only three copies of the original are known to survive.[8]

✧   ✧   ✧

Mercator died in Duisburg on December 2, 1594, having lived a long life that brought him fame and wealth. Yet his most famous achievement, the map that bears his name, was not immediately embraced by the maritime community, who could not understand its excessive distortion of the shape of continents. The fact that Mercator had not given a full account of how he had "progressively increased" the distance between the parallels only added to the confusion. It was left to Edward Wright (ca. 1560–1615), an English mathematician and instrument maker, to give the first accurate account of the principles underlying Mercator's map. In a work entitled *Certaine Errors in Navigation . . .* , published in London in 1599, he wrote:

The parts of the meridians at euery poynt of latitude must increase with the same proportion wherewith the Secantes increase. By perpetuall addition of the Secantes answerable to the latitude of each parallel vnto the summe compounded of all former secantes . . . we may make a table which shall truly shew the points of latitude in the meridians of the nautical planisphaere.[9]

In other words, Wright used *numerical integration* to evaluate $\int_0^\phi \sec\phi\,d\phi$. Let us follow his plan, using modern notation.

Figure 85 shows a small spherical rectangle defined by the circles of longitudes λ and $\lambda + \Delta\lambda$ and circles of latitudes ϕ and $\phi + \Delta\phi$, where λ and ϕ are measured in radians. (Because the choice of the "zero meridian" is arbitrary, only the difference in longitude $\Delta\lambda$ is shown in the figure.) The sides of this rectangle have length $(R\cos\phi)\,\Delta\lambda$ and $R\,\Delta\phi$, respectively. Let a point $P(\lambda, \phi)$ on the sphere go over to the point $P'(x, y)$ on the map (where $y = 0$ corresponds to the equator). Then the spherical rectangle will be mapped onto a planar rectangle defined by the lines $x, x + \Delta x, y,$ and $y + \Delta y$, where $\Delta x = R\,\Delta\lambda$. Now, the requirement that the map be conformal means that these two rectangles must be *similar* (which in turn means that the direction from $P(\lambda, \phi)$ to a neighboring point $Q(\lambda + \Delta\lambda, \phi + \Delta\phi)$ is the same as between their images on the map). Thus we are led to the equation

$$\frac{\Delta y}{R\,\Delta\lambda} = \frac{R\,\Delta\phi}{R\cos\phi\,\Delta\lambda}$$

or

$$\Delta y = (R\sec\phi)\,\Delta\phi. \qquad (3)$$

In modern terms, equation (3) is a *finite difference* equation. It can be solved numerically by a step-by-step procedure: we put

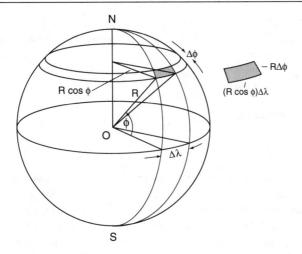

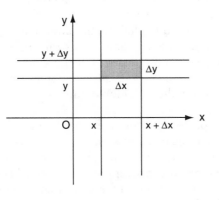

FIG. 85.
Spherical
rectangle on
the globe and
its projection
on Mercator's
map.

$\Delta y_i = y_i - y_{i-1}$, $i = 1, 2, 3, \ldots$, and decide on a fixed increment $\Delta\phi$. Starting with the equator ($y_0 = 0$), we increase ϕ by $\Delta\phi$, find Δy_1 from equation (3), and then find $y_1 = y_0 + \Delta y_1$; we again increase ϕ by $\Delta\phi$ and find $y_2 = y_1 + \Delta y_2$, and so on until the desired range of latitudes has been covered. This *numerical integration* is a tedious, time-consuming procedure—unless one has a programmable calculator or computer, neither of which was available to Wright. Nevertheless he carried the plan through, continually adding the secants at intervals of one minute of arc.[10] He published his results in a table of "meridional parts" for latitudes from 0° to 75°. So at last the method of construction of Mercator's map became known.

Nowadays, of course, we would write equation (3) as a *differential equation*: we let both $\Delta\phi$ and Δy become infinitely small, and in the limit get

$$\frac{dy}{d\phi} = R \sec \phi, \tag{4}$$

whose solution is

$$y = R \int_0^\phi \sec t \, dt \tag{5}$$

(we have used t instead of ϕ in the integrand to distinguish the variable of integration from the upper limit of the integral). Today this integral is given as an exercise in a second-semester calculus class (we will say more about it shortly). But Wright's book appeared some seventy years before Newton and Leibniz invented the calculus, so he could not avail himself of its techniques. He had no choice but to resort to numerical integration.

Being a scholar, Wright wrote his book for readers versed in mathematics. But to the ordinary mariner such theoretical explanations meant very little. So Wright devised a simple physical model that he hoped would explain to the uninitiated the principles behind Mercator's map: imagine we wrap the globe in a cylinder touching it along the equator. Let the globe "swell like a bladder" so that each point on its surface comes into contact with the cylinder. When the cylinder is unwrapped, you get a Mercator map.

Unfortunately for posterity—and due to no fault of Wright's—this descriptive model became the source of a persisting myth: that Mercator's map is obtained by projecting rays of light from the center of the globe to the wrapping cylinder (this actually produces the cylindrical projection we discussed earlier in this chapter). Technically, Mercator's "projection" is not a projection at all, at least not in the geometric sense of the word: it can only be obtained by a mathematical procedure that, at its core, involves an infinitesimal process and hence the calculus. Mercator himself never used the cylinder concept, and his projection—except for a superficial similarity—has nothing to do with the cylindrical projection. But once created, myths are slow to die, and even today one finds erroneous statements to that effect in many geography textbooks.

Additional misunderstandings came from the map's excessive distortion of lands at high latitudes: Greenland, for example, appears larger than South America, though in reality it is only one-ninth as big. Moreover, a straight-line segment connecting two points on the map does *not* represent the shortest distance

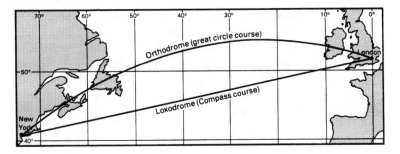

FIG. 86. Rhumb line (loxodrome) and great-circle arc on Mercator's projection.

between them on the globe (unless both points are on the equator or on the same meridian), as figure 86 shows. These "shortcomings" have often been used to criticize Mercator's map. An anonymous contemporary of Wright, evidently offended by this unjust criticism, vented his frustration in these lines:

> *Let no one dare to attribute the shame*
> *of misuse of projections to Mercator's name;*
> *but smother quite, and let infamy light*
> *upon those who do misuse, publish or recite.*

As we have seen, Mercator himself had already realized that no single map could at once preserve distance, shape, *and* direction. Having the navigator's need in mind, he chose to sacrifice distance and shape in order to preserve direction. Yet many people's perception of the world still comes from the large Mercator map that hung from the wall of their high school classroom.

✧   ✧   ✧

Wright's book appeared in 1599, thirty years after Mercator published his new world map. Slowly the maritime community began to appreciate the great value of the map to navigators, and in due time it became the standard map for global navigation, a status which it has kept ever since. When NASA began its space exploration in the 1960s, a huge Mercator map, on which the trajectories of satellites were being continuously monitored, dominated the Mission Control room in Houston, Texas. And the first maps of the moons of Jupiter and Saturn, photographed from close range by the Pioneer and Voyager spacecraft, were drawn on his projection.

But let us return to the seventeenth century, where the story now shifts to the mathematical arena. In 1614 John Napier (1550–1617) of Scotland published his invention of logarithms, the single most important aid to computational mathematics since the Hindu-Arabic numeration system was brought to Europe in the Middle Ages.[11] Shortly thereafter Edmund Gunter (1581–1626), an English mathematician and clergyman, published a table of logarithmic tangents (1620). Around 1645 Henry Bond, a mathematics teacher and authority on navigation, compared this table with Wright's meridional table and noticed to his surprise that the two tables matched, provided the entries in Gunter's table were written as $(45° + \phi/2)$. He conjectured that $\int_0^\phi \sec t \, dt$ *is equal to* $\ln \tan (45° + \phi/2)$, where "ln" stands for natural logarithms (logarithms to the base $e = 2.718\ldots$), but he could not prove it. Soon his conjecture became one of the outstanding mathematical problems of the 1650s. It was unsuccessfully attempted by John Collins, Nicolaus Mercator (no relation to Gerhard), Edmond Halley (of comet fame), and others—all contemporaries of Isaac Newton and active participants in the developments that led to the invention of the calculus.

Finally in 1668 James Gregory, whom we have already met in connection with the Gregory-Leibniz series, succeeded in proving Bond's conjecture; his proof, however, was so difficult that Halley denounced it as being full of "complications." So it befell Isaac Barrow (1630–1677), who preceded Newton as the Lucasian Professor of Mathematics at Cambridge University, to give an "intelligent" proof of Bond's conjecture (1670). And in so doing he seems to have been the first to use the technique of decomposition into partial fractions, so effective in solving numerous indefinite integrals. The details of his proof are given in Appendix 2.

✧ ✧ ✧

We are now in a position to write down the coordinates (x, y) of a point P' on Mercator's Map in terms of longitude λ and latitude ϕ of the corresponding point P on the globe. The difference equation $\Delta x = R \Delta \lambda$ (p. 174) has the obvious solution $x = R\lambda$, and the integral appearing in equation (5) is equal to $\ln \tan (45° + \phi/2)$; we thus have

$$x = R\lambda, \quad y = R \ln \tan (45° + \phi/2).^{12} \tag{6}$$

Figure 87 shows the world as it appears on Mercator's map; because of the excessive north-south stretching at high latitudes, the map is confined to latitudes from 75° north to 60° south.

But our story does not quite end yet. The reader may have noticed that the expression $\tan (45° + \phi/2)$ inside the logarithm

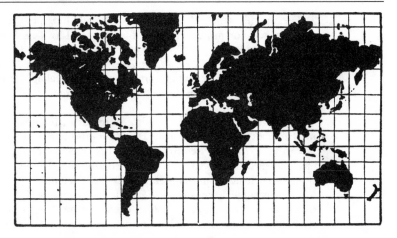

FIG. 87. The world on Mercator's map.

in equation (6) is the same as that appearing in equation (2) in connection with the stereographic projection. This is no coincidence. One of the great achievements of eighteenth-century mathematics was to extend the algebra of ordinary functions such as $\sin x$, e^x, and $\ln x$ to *imaginary* and even *complex* values of the variable x. This development began with Euler and reached its climax in the nineteenth century with the theory of functions of a complex variable. As we shall see in the next chapter, this extension enables us to regard Mercator's projection as a conformal *mapping* (in the mathematical as well as geographical sense) of the stereographic projection by means of the function $w = \ln u$, where both u and w are complex variables.

NOTES AND SOURCES

1. A detailed discussion of the stereographic projection and its relation to inversion can be found in my book *To Infinity and Beyond: A Cultural History of the Infinite* (1987: rpt. Princeton, N.J.: Princeton University Press, 1991), pp. 95–98 and 239–245. For other map projections, see Charles H. Deetz and Oscar S. Adams, *Elements of Map Projection* (New York: Greenwood Press, 1969), and John P. Snyder, *Flattening the Earth: Two Thousand Years of Map Projections* (Chicago: University of Chicago Press, 1993).

2. From the Greek *loxos* = slanted, and *dromos* = course, i.e., a slanted line. The term was coined in 1624 by the Dutch scientist Willebrord van Roijen Snell (1581–1626), known for his law of refraction in optics.

3. Despite his fame, there is no full-length biography of Mercator in English. For more details on his life, see Lloyd A. Brown, *The Story of Maps* (1949; rpt. New York: Dover, 1977), pp. 134–136 and 158–160; Robert W. Karrow, *Mapmakers of the Sixteenth Century and Their Maps* (Chicago: Speculum Orbis Press, 1993), chap. 56; and John Noble Wilford, *The Mapmakers* (New York: Alfred A. Knopf, 1981), pp. 73–77.

4. The priority for this idea goes to Mercator's contemporary, Abraham Ortelius (1527–1598), whose *Theatrum orbis terrarum* appeared in Antwerp in 1570 and is considered the first modern atlas. The word "atlas," however, is due to Mercator. Mercator and Ortelius were competitors in their field but maintained a friendly relation.

5. On this subject see Snyder, *Flattening the Earth*, p. 47.

6. Mercator soon realized that he could not meet both of these requirements: his map could not preserve direction *and* distance, so he dropped the distance requirement.

7. This version is taken from the article "An Application of Geography to Mathematics: History of the Integral of the Secant" by V. Frederick Rickey and Philip M. Tuchinsky, the *Mathematics Magazine*, vol. 53, no. 3 (May 1980). A slightly different version appears in Snyder, *Flattening the Earth*, pp. 46–47.

8. Their locations are listed in R. V. Tooley, *Maps and Map-Makers* (New York: Bonanza Books, 1962), p. 31; the copy listed in Breslau, Germany was destroyed in World War II.

9. Rickey and Tuchinsky, "Application of Geography," (see again Snyder, *Flattening the Earth*, p. 48, for a slightly different version). For ease of reading I have slightly edited the original text, omitting repetetive phrases but not affecting the context. The full title of Wright's book is *Certaine errors in navigation, arising either of the ordinarie erroneous making or vsing of the sea chart, compasse, crosse staffe, and tables of declination of the sunne, and fixed starres detected and corrected* (long titles were necessary in those days to attract the readers' attention). Wright was a fellow at Caius College, Cambridge, and became tutor to Henry, Prince of Wales, the son of King James I. He is best known for his English translation of Napier's work of logarithms.

10. Snyder, *Flattening the Earth*, p. 48. Florian Cajori, in *A History of Mathematics* (1893; 2d ed. New York: Macmillan, 1919), p. 189, claims that Wright used intervals of one arc *second*; this, however, seems unlikely, for it would imply $3600 \times 75 = 270,000$ additions in order to cover the latitude range from $0°$ to $75°$.

11. On the history of logarithms, see my book, *e: The Story of a Number* (Princeton, N.J.: Princeton University Press, 1994), chaps. 1 and 2.

12. An equivalent form of the second of equations (6) is $y = R \ln (\sec \phi + \tan \phi)$.

14

---

sin *x* = 2: Imaginary Trigonometry

De Moivre's theorem was the key to a whole new world
of imaginary or complex trigonometry.
—Herbert Mc Kay, *The World of Numbers*
(1946), p. 157

Imagine you have just bought a brand new hand-held calculator and find to your dismay that when you try to subtract 5 from 4, you get an error sign. Yet this is precisely the situation in which first-grade pupils would find themselves if their teacher asked them to take away five apples from four: "It cannot be done!" would be the class's predictable response.

The history of mathematics is full of attempts to break the barrier of the "impossible." Many of these attempts have ended in failure: for more than two thousand years mathematicians tried to find a construction, using straightedge and compass alone, that would trisect an arbitrary angle—until it was proved, around the middle of the nineteenth century, that such a construction is impossible. The countless attempts to "square the circle"—to construct, again with straightedge and compass alone, a square whose area equals that of a given circle—have likewise proved futile (which does not prevent amateurs from submitting hundreds of proposed "solutions" for publication, to the dismay of editors of mathematics journals).

But there have also been brilliant successes: the acceptance of negative numbers into mathematics has freed arithmetic from the interpretation of subtraction as an act of "taking away," with the consequence that a whole new range of problems could be considered—from financial problems (credit and debit) to solving general linear equations. And breaking the taboo—so deeply rooted in our mathematical instincts—against dealing with the square root of a negative number paved the way to the algebra of imaginary and complex numbers, culminating in the powerful theory of functions of a complex variable. The story of these extensions of our number system, with their many false turns and ultimate triumph, has been told elsewhere;[1] here we are mainly concerned with its implications to trigonometry.

One of the first things we learn in trigonometry is that the domain of the function $y = \sin x$ is the set of all real numbers, and its range the interval $-1 \le y \le 1$; accordingly, if you try to find an angle whose sine is, say, 2, your calculator, after pressing ARCSIN (or SIN^{-1}, or INV SIN), will show an error sign—just as most calculators would do when trying to find $\sqrt{-1}$. Yet at the beginning of the eighteenth century, attempts were made to extend the function concept to imaginary and even complex values of the independent variable; these attempts would prove enormously successful; among other things, they enable us to solve the equation $\sin x = y$ when y has *any* given value—real, imaginary, or complex.

One of the earliest pioneers in this direction was Roger Cotes (1682–1716). In 1714 he published the formula

$$i\phi = \log (\cos \phi + i \sin \phi),$$

where $i = \sqrt{-1}$ and "log" means natural logarithm; it was reprinted in his only major work, *Harmonia mensurarum*, a compilation of his papers published posthumously in 1722. Cotes worked on a wide range of problems in mathematics and astronomy (see p. 82) and served as editor of the second edition of Newton's *Principia*, but his sudden death at the age of thirty-four prematurely ended his promising career; Newton said of him: "Had Cotes lived we might have known something."[2]

No doubt because of his early death, Cotes never got the credit due to him for discovering this ground-breaking formula; it is instead named after Euler and written in reverse,

$$e^{i\phi} = \cos \phi + i \sin \phi. \tag{1}$$

In this form it appeared in Euler's great work, *Analysin infinitorum* (1748), together with the companion formula,

$$e^{-i\phi} = \cos (-\phi) + i \sin (-\phi)$$
$$= \cos \phi - i \sin \phi.$$

By adding and subtracting these formulas Euler obtained the following expressions for $\cos \phi$ and $\sin \phi$:

$$\cos \phi = \frac{e^{i\phi} + e^{-i\phi}}{2}, \sin \phi = \frac{e^{i\phi} - e^{-i\phi}}{2i}. \tag{2}$$

These two formulas are the basis of modern analytic trigonometry.

The credit given to Euler for rediscovering Cotes's formula is not entirely undeserved: whereas Cotes (as also his contemporary De Moivre) was still treating complex numbers merely as a convenient, if mysterious, way of shortening algebraic computations, it was Euler who fully incorporated these numbers into the algebra of *functions*. His idea was that a complex number can be used as an input to a function, *provided the output is also a complex number*.

Take, for example, the function $w = \sin z$, where both z and w are complex variables. Writing $z = x + iy$, $w = u + iv$ and proceeding as if the laws of ordinary trigonometry still hold, we get

$$w = u + iv = \sin(x + iy)$$

$$= \sin x \cos iy + \cos x \sin iy. \tag{3}$$

But what *are* $\cos iy$ and $\sin iy$? Again proceeding in a purely formal way, let us substitute iy for ϕ in equations (2):

$$\cos iy = \frac{e^{i(iy)} + e^{-i(iy)}}{2} = \frac{e^y + e^{-y}}{2},$$

$$\sin iy = \frac{e^{i(iy)} - e^{-i(iy)}}{2i} = \frac{e^{-y} - e^y}{2i} = \frac{i(e^y - e^{-y})}{2}.$$

It so happens that the expressions $(e^y + e^{-y})/2$ and $(e^y - e^{-y})/2$ exhibit many formal similarities with the functions $\cos y$ and $\sin y$, respectively, and are consequently denoted by $\cosh y$ and $\sinh y$ (read "hyperbolic cosine" and "hyperbolic sine" of y):

$$\cosh y = \frac{e^y + e^{-y}}{2}, \sinh y = \frac{e^y - e^{-y}}{2}. \tag{4}$$

For example, if we square the two expressions and subtract the results, we get the identity

$$\cosh^2 y - \sinh^2 y = 1, \tag{5}$$

in analogy with the trigonometric identity $\cos^2 y + \sin^2 y = 1$ (note, however, the minus sign of the second term). We also have $\cosh 0 = 1$, $\sinh 0 = 0$, $\cosh(-y) = \cosh y$, $\sinh(-y) = -\sinh y$, $\cosh(x \pm y) = \cosh x \cosh y \pm \sinh x \sinh y$, $\sinh(x \pm y) = \sinh x \cosh y \pm \cosh x \sinh y$, and $d(\cosh y)/dy = \sinh y$, $d(\sinh y)/dy = \cosh y$. It turns out that most of the familiar trigonometric formulas have their hyperbolic counterparts, with a possible change of sign in one of the terms.[3]

We can now write equation (3) as

$$\sin z = \sin x \cosh y + i \cos x \sinh y, \tag{6}$$

where again $z = x + iy$. In exactly the same way we can find an expression for $\cos z$:

$$\cos z = \cos x \cosh y - i \sin x \sinh y. \tag{7}$$

As an example, let us find the sine of the complex number $z = 3 + 4i$. Taking all units in radians, we have $\sin 3 = 0.141$, $\cosh 4 = 27.308$, $\cos 3 = -0.990$ and $\sinh 4 = 27.290$ (all rounded to three places), so $\sin z = \sin 3 \cosh 4 + i \cos 3 \sinh 4 = 3.854 - 27.017i$.

There is, of course, one serious flaw in what we have just done: the very assumption that the familiar rules of algebra and trigonometry of real numbers still hold when applied to complex numbers. There is really no a priori guarantee that this should be the case; indeed, occasionally these rules break down when extended beyond their original domain: we may not, for example, use the rule $\sqrt{a} \cdot \sqrt{b} = \sqrt{ab}$ when a and b are negative, for otherwise we would have $i^2 = (\sqrt{-1}) \cdot (\sqrt{-1}) = \sqrt{[(-1) \cdot (-1)]} = \sqrt{1} = 1$, instead of -1. But these subtleties did not prevent Euler from playing with his new idea: he lived in an era when a carefree manipulation of symbols was still accepted, and he made the most of it. He simply had faith in his formulas, and usually he was right. His daring and imaginative explorations resulted in numerous new relations whose rigorous proof had to await future generations.

Of course, an act of faith is not always a reliable guide in science, least of all in mathematics. The theory of functions of a complex variable, or *theory of functions*, as it is known for short, was created in part to put Euler's ideas on a firm ground. It does this by essentially turning the tables around: we *define* a function $w = f(z)$ in such a way that all the properties of the *real-valued* function $y = f(x)$ will be preserved when x and y are replaced by the complex variables z and w. Moreover, we should always be able to get back the "old" function $f(x)$ as a special case of the "new" function when z is a real variable x (that is, when $z = x + oi$).

Let us illustrate these ideas with the sine and cosine functions. Using equations (6) and (7) as the definitions of $\sin z$ and $\cos z$, we can show that $\sin^2 z + \cos^2 z = 1$, that each has a period 2π (that is, $\sin(z + 2\pi) = \sin z$ for all z, and similarly for $\cos z$), and that the familiar addition formulas still hold. Under certain conditions we can even differentiate a complex-valued function $f(z)$, giving us (formally) the same derivative as we would get when differentiating the real-valued function $y = f(x)$;[4] in our case we have $d(\sin z)/dz = \cos z$ and $d(\cos z)/dz = -\sin z$, exactly as in the real case.

But, you may ask, if the extension of a real-valued function to the complex domain merely reproduces its old properties, why go through the trouble? It would certainly not be worth the effort, were it not for the fact that this extension endows the function with some new properties unique to the complex domain. Foremost among these is the concept of a *mapping* from one plane to another.

To see this, we must first reexamine the function concept when applied to a complex variable. A real-valued function $y = f(x)$ assigns to every real number x (the "input" or "independent variable") in its domain one, and only one real number y (the "output" or "dependent variable") in the range; it is thus a "mapping" from the x-axis to the y-axis. A convenient way to depict this mapping is to graph the function in the xy coordinate plane—in essence producing a pictorial representation that lets us see, quite literally, the manner in which the two variables depend on each other.

However, when we try to extend this idea to complex variables—that is, replace the real-valued function $y = f(x)$ with the complex-valued function $w = f(z)$—we immediately encounter a difficulty. To plot a single complex number $x + iy$ requires a two-dimensional coordinate system—one coordinate for the real part x and another for the imaginary part y. But now we are dealing with *two* complex variables, z and w, each of which requires its own two-dimensional coordinate system. We cannot therefore "graph" the function $w = f(z)$ in the same sense as we graph the function $y = f(x)$; to describe it geometrically, we need to think of it as *a mapping from one plane to another*.

Let us illustrate this with the function $w = z^2$, where $z = x + iy$ and $w = u + iv$. We have

$$w = u + iv = (x + iy)^2$$
$$= x^2 + 2ixy + (iy)^2 = (x^2 - y^2) + (2xy)i.$$

Equating real and imaginary parts, we get

$$u = x^2 - y^2, v = 2xy. \tag{8}$$

Equations (8) tell us that both u and v are functions of the two independent variables x and y. Let us call the xy-plane the "z-plane" and the uv-plane the "w-plane." Then the function $w = z^2$ maps every point $P(x, y)$ of the z-plane onto a corresponding point $P'(u, v)$, the image of P in the w-plane; for example, the point $P(3, 4)$ goes over to the point $P'(-7, 24)$, as follows from the equation $(3 + 4i)^2 = -7 + 24i$.

Imagine now that P describes some curve in the z-plane; then P' will describe an "image curve" in the w-plane. For example, if P moves along the equilateral hyperbola $x^2 - y^2 = $ constant, P' will move along the curve $u = $ constant, which is a vertical line in the w-plane. Similarly, if P traces the equilateral hyperbola $2xy = $ constant, its image will trace the horizontal line $v = $ constant. By assigning different values to the constants, we get two *families* of hyperbolas in the z-plane; their images form a rectangular grid in the w-plane (fig. 88).

One of the most elegant results in the theory of functions says that the mapping effected by a function $w = f(z)$ is *conformal* (direction-preserving) at all points z where it has a non-zero derivative.[5] This means that if two curves in the z-plane intersect at a certain angle (the angle between their tangent lines at the point of intersection), their image curves in the w-plane will intersect at the same angle, provided $df(z)/dz$ exists and is not zero at the point of intersection. This is clearly seen in the case of $w = z^2$: the two families of hyperbolas, $x^2 - y^2 = $ constant and $2xy = $ constant, are *orthogonal*—each hyperbola of one family intersects every hyperbola of the other at a right angle—as are their image curves, the horizontal and veritcal lines in the w-plane.

The mapping effected by the function $w = \sin z$ can be explored in a similar way. We start with horizontal lines $y = c = $ constant in the z-plane. Equations (6) then tell us that

$$u = \sin x \cosh c, \quad v = \cos x \sinh c. \tag{9}$$

Equations (9) can be regarded as a pair of parametric equations describing a curve in the w-plane, the parameter being x. To

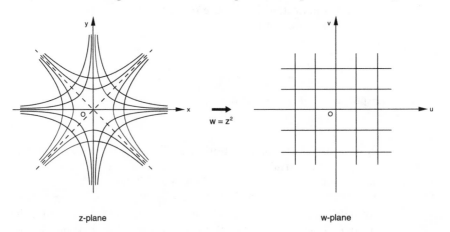

z-plane $w = z^2$ w-plane

FIG. 88. Mapping by the function $w = z^2$.

get the rectangular equation of this curve, we must eliminate x between the two equations. We can do this by dividing the first equation by $\cosh c$ and the second by $\sinh c$, squaring the results and adding; in view of the identity $\sin^2 x + \cos^2 x = 1$ we get

$$\frac{u^2}{\cosh^2 c} + \frac{v^2}{\sinh^2 c} = 1. \tag{10}$$

Equation (10) is of the form $u^2/a^2 + v^2/b^2 = 1$, which represents an ellipse with center at the origin of the w-plane, semimajor axis $a = \cosh c$ and semiminor axis $b = |\sinh c|$ (since $\cosh y$ is always greater than $\sinh y$—as follows from equations 4— the major axis is always along the u-axis). Note, however, that it takes the *pair* of lines $y = \pm c$ to produce the entire ellipse, the upper half corresponding to $y = c$ (where $c > 0$), the lower half to $y = -c$ (this can best be seen from the parametric equations 9). From analytic geometry we know that the two *foci* of the ellipse are located at the points $(\pm f, 0)$, where $f^2 = a^2 - b^2$. But $a^2 - b^2 = \cosh^2 c - \sinh^2 c = 1$, so we have $f = \pm 1$; for different values of c we therefore get a family of ellipses with a common pair of foci at $(\pm 1, 0)$, regardless of c. As $c \to 0$, $\cosh c \to 1$ and $\sinh c \to 0$, so that the ellipses gradually narrow until they degenerate into the line segment $-1 \leq u \leq 1$ along the u-axis. These features are shown in figure 89.

Next consider vertical lines $x = k = $ constant in the z-plane. Equations (9) give us

$$u = \sin k \cosh y, \quad v = \cos k \sinh y. \tag{11}$$

This time we can eliminate the parameter y by dividing the first equation by $\cosh y$ and the second by $\sinh y$, squaring the results and subtracting; in view of the identity $\cosh^2 y - \sinh^2 y = 1$ we

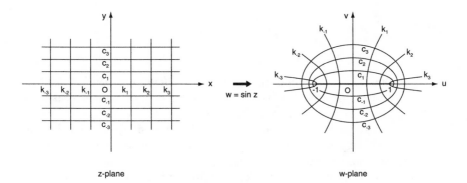

z-plane   w-plane

FIG. 89. Mapping by the function $w = \sin z$.

get

$$\frac{u^2}{\sin^2 k} - \frac{v^2}{\cos^2 k} = 1. \tag{12}$$

Equation (12) is of the form $u^2/a^2 - v^2/b^2 = 1$, whose graph is a hyperbola with center at the origin, semi-transverse axis $a = |\sin k|$ and semi-conjugate axis $b = |\cos k|$; its asymptotes are the pair of lines $y = \pm[(\cos k)/(\sin k)]x = \pm(\cot k)x$. As with the ellipses, it takes the *pair* of lines $x = \pm k$ to produce the entire hyperbola, the right branch corresponding to $x = k$ (where $k > 0$), the left branch to $x = -k$. The two foci of the hyperbola are located at $(\pm f, 0)$, where now $f^2 = a^2 + b^2 = \sin^2 k + \cos^2 k = 1$; changing the value of k therefore produces a family of hyperbolas with a common pair of foci at $(\pm 1, 0)$ (see fig. 89). As $k \to 0$ the hyperbolas open up, and for $k = 0$ (corresponding to the x-axis in the z-plane) they degenerate into the line $u = 0$ (the v-axis in the w-plane). On the other hand, as $|k|$ *increases* the hyperbolas narrow, degenerating into the pair of rays $u \geq 1$ and $u \leq -1$ when $k = \pm \pi/2$. We also note that any increase of k by π does not change the values of $\sin^2 k$ or $\cos^2 k$ and so produces the same hyperbola; this, of course, merely shows that the mapping is not one-to-one, as we already know from the periodicity of $\sin z$. Finally, the ellipses and hyperbolas form orthogonal families, as follows from the conformal property of complex-valued functions.[6]

✧   ✧   ✧

As one more example we consider the function $w = e^z$. First, of course, we must define what we mean by e^z, so let us proceed formally by writing $z = x + iy$ and acting as if the rules of algebra of real numbers are still valid:

$$e^z = e^{x+iy} = e^x e^{iy}.$$

But $e^{iy} = \cos y + i \sin y$, so we have

$$e^z = e^x(\cos y + i \sin y). \tag{13}$$

We now regard equation (13) as the *definition* of e^z. We note, first of all, that this is not a one-to-one function: increasing y by 2π does not change the value of e^z, so we have $e^{z+2\pi i} = e^z$. Consequently, the complex-valued exponential function has an *imaginary period* $2\pi i$. This, of course, is in marked contrast to the real-valued function e^x.

Writing $w = e^z = u + iv$, equation (13) implies that

$$u = e^x \cos y, \, v = e^x \sin y. \tag{14}$$

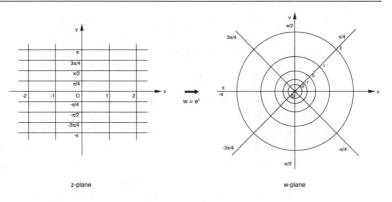

FIG. 90. Mapping by the function $w = e^z$.

Putting $y = c =$ constant in these equations and eliminating the parameter x, we get $v = (\tan c)u$, the equation of a ray of slope $\tan c$ emanating from the origin of the w-plane; putting $x = k =$ constant and eliminating y, we get $u^2 + v^2 = e^{2k}$, representing a circle with center at the origin and radius e^k. Thus the rectangular coordinate grid of the z-plane is mapped on a *polar* grid in the w-plane, with the circles spaced exponentially (fig. 90). Again the two grid systems are everywhere orthogonal.

We may discover here an unexpected connection with the map projections discussed in the previous chapter. To produce a polar grid in the w-plane with circles spaced *linearly* (rather than exponentially), we must increase k *logarithmically*; that is, vertical lines in the z-plane must be spaced as $\ln k$. The complex-valued function that accomplishes this is the *inverse* of $w = e^z$, that is, $w = \ln z$. This mapping, when applied to the polar grid of a stereographic projection, gives us none other than Mercator's grid—except that the roles of horizontal and vertical lines are reversed. But this reversal can be corrected by a 90° rotation of the coordinate system, that is, by a multiplication by i. We can therefore start with the globe, project it stereographically onto the z-plane, and then map this plane onto the w-plane by means of the function $w = i \ln z$: the final product is Mercator's projection. And since each of the component maps is conformal, so will be their product. As we recall, it was precisely this goal—to produce a conformal (direction-preserving) world map based on a rectangular grid—that led Mercator to his famous projection.

✧ ✧ ✧

We began this chapter with the observation that no real angle exists whose sine is 2. But now that we have extended the functions $\sin x$ and $\cos x$ to the complex domain, let us try again. We wish to find an "angle" $z = x + iy$ such that $\sin z = 2$. From equations (6) it follows that

$$\sin x \cosh y = 2, \cos x \sinh y = 0.$$

The second of these equations implies that $\sinh y = 0$ or $\cos x = 0$; that is, $y = 0$ or $x = (n + 1/2)\pi$, $n = 0, \pm 1, \pm 2, \dots$. Putting $y = 0$ in the first equation gives us $(\sin x)(\cosh 0) = \sin x = 2$, which has no solution because x is a real number. Putting $x = (n + 1/2)\pi$ in the first equation gives us $[\sin (n + 1/2)\pi](\cosh y) = (-1)^n \cosh y = 2$, so that $\cosh y = \pm 2$. But the range of $\cosh y$ is $[1, \infty)$, as follows from the defining equation $\cosh y = (e^y + e^{-y})/2$; hence we only need to consider $\cosh y = 2$. Since $\cosh y$ is an even function, this equation has two equal but opposite solutions; these can be found from a table or a calculator that has hyperbolic-function capabilities: we find $y = \pm 1.317$, rounded to three places. Thus the equation $\sin z = 2$ has infinitely many solutions $z = (n + 1/2)\pi \pm 1.317i$, where $n = 0, \pm 1, \pm 2, \dots$. None of these solutions is a real number.

All of this may seem quite abstract and removed from ordinary trigonometry; it is certainly strange to talk of imaginary angles and their sines. Yet strangeness is a relative concept; with sufficient familiarity, the "strange" of yesterday becomes the commonplace of today. When negative numbers began to appear on the mathematical scene, they were at first regarded as strange, artificial creations ("how can one subtract five objects from four?"). A similar reaction awaited imaginary numbers, as attested by the very name "imaginary." When Euler pioneered the extension of ordinary functions to the complex domain, his bold conclusions were strange and controversial; he was the first, for example, to define the logarithm of a negative number in terms of imaginary numbers—this at a time when even the existence of imaginary numbers was not yet entirely accepted.

It took the authority of Gauss to fully incorporate complex numbers into algebra; this he did in 1799 in his doctoral dissertation at the age of twenty-one, in which he gave the first rigorous demonstration of the *fundamental theorem of algebra*: a polynomial of degree n has exactly n (not necessarily different) roots in the complex number system.[7] And any lingering doubts as to the "existence" of these numbers were put to rest when Sir William Rowan Hamilton (1805–1865) in 1835 presented his elegant definition of complex numbers as pairs of real numbers subject to a formal set of rules.[8] The door was now open to a vast expansion of the methods of analysis to complex variables, culminating in the *theory of functions* with its numerous applications in nearly every branch of mathematics, pure or applied. The "strange" of yesterday indeed became the commonplace of today.

NOTES AND SOURCES

1. On the history of negative numbers, see David Eugene Smith, *History of Mathematics* (1925; rpt. New York: Dover, 1958), vol. 2, pp. 257–260; on the history of imaginary and complex numbers, see ibid., pp. 261–268.

2. More on Cotes's life and work can be found in Stuart Hollingdale, *Makers of Mathematics* (Harmondsworth, U.K.: Penguin Books, 1989), pp. 245–252.

3. Note, however, that the hyperbolic functions are *not* periodic and that their ranges are $1 \leq \cosh y < \infty$ and $-\infty < \sinh y < \infty$. The name "hyperbolic" comes from the fact that if we write $x = \cosh t$, $y = \sinh t$ (where t is a real parameter), then the identity $\cosh^2 t - \sinh^2 t = 1$ implies that a point with coordinates (x, y) lies on the equilateral hyperbola $x^2 - y^2 = 1$, just as a point with coordinates $x = \cos t$, $y = \sin t$ lies on the unit circle $x^2 + y^2 = 1$. For a history of the hyperbolic functions, see my book, *e: The Story of a Number* (Princeton, N.J.: Princeton University Press, 1994), pp. 140–150 and 208–210.

4. However, the concept of a derivative of a complex-valued function involves some subtleties that are not present in the real domain. An alternative approach, due to the German mathematician Karl Weierstrass (1815–1897), is to define a function by means of a power series, in the case of $\sin z$ the series $z - z^3/3! + z^5/5! + - \cdots$. For details, see any book on the theory of functions.

5. This is a far-reaching result which we have stated here only in brief form; for the complete theorem, see any book on the theory of functions.

6. For a more detailed discussion of the mapping $w = \sin z$, see Erwing Kreiszig, *Advanced Engineering Mathematics* (New York: John Wiley, 1979), pp. 619–620.

7. For example, the polynomial $f(z) = z^3 - 1$ has the three roots 1, $(-1 - i\sqrt{3})/2$, and $(-1 + i\sqrt{3})/2$, as can be seen by factoring $f(z)$ into $(z - 1)(z^2 + z + 1)$, setting each factor equal to zero and solving the resulting equations. These "cubic roots of unity" can be written in trigonometric form as cis 0, cis $2\pi/3$, and cis $4\pi/3$, where "cis" stands for $\cos + i \sin$. It always surprises students to learn that the number 1 has *three* cubic roots, two of which are complex.

8. See Maor, *e: The Story of a Number*, pp. 166–168.

Edmund Landau: The Master Rigorist

Edmund Yehezkel Landau was born in Berlin in 1877; his father was the well-known gynecologist Leopold Landau. He began his education at the French Gymnasium (high school) in Berlin and soon thereafter devoted himself entirely to mathematics. Among his teachers was Ferdinand Lindemann (1852–1939), who in 1882 proved the transcendence of π—the fact that π cannot be the root of a polynomial equation with integer coefficients—thereby settling the age-old problem of "squaring the circle" (see p. 181). From the beginning Landau was interested in analytic number theory—the application of analytic methods to the study of integers. In 1903 he gave a simplified proof of the Prime Number Theorem, which had first been proved seven years earlier after having defied some of the greatest minds of the nineteenth century.[1] In 1909, when he was only 31, Landau was appointed professor of mathematics at the University of Göttingen, the world-renowned center of mathematical research up until World War II; he succeeded Hermann Minkowski (1864–1909), known for his four-dimensional interpretation of Einstein's theory of relativity, who had died at the age of forty-five. Landau published over 250 papers and wrote several major works in his field, among them *Handbook of the Theory and Distribution of the Prime Numbers* (in two volumes, 1909) and *Lectures on Number Theory* (in three volumes, 1927).

Landau was one of eight distinguished scholars who were invited to talk at the ceremonies inaugurating the Hebrew University of Jerusalem in 1925. From atop Mount Scopus overlooking the Holy City, he spoke about "Solved and Unsolved Problems in Elementary Number Theory"—a rather unusual subject for such a festive occasion. He accepted the university's invitation to occupy its first chair of mathematics, teaching himself Hebrew expressly for this purpose. He joined the university in 1927 but shortly afterward returned to Germany to resume his duties at Göttingen. His brilliant career, however, was soon to come to an end: when the Nazis came to power in 1933 he—along with all Jewish professors at German universities—was forced to resign his position. His sudden death in 1938 saved him from the fate awaiting the Jewish community in Germany.

Landau embodied the ultimate image of the pure mathematician. He viewed any practical applications to mathematics with disdain and avoided the slightest reference to them, dismissing them as *Schmieröl* (grease); among the "practical applications" was geometry, which he entirely shunned from his exposition. In his lectures and written work, definitions, theorems, and proofs followed in quick succession, without the slightest hint at the motivation behind them. His goal was absolute and uncompromising rigor. His assistant, who always attended his lectures, was instructed to interrupt him if the professor omitted the slightest detail.[2]

To the student of higher mathematics Landau is best known for his two textbooks, *Grundlagen der Analysis* (Foundations of analysis, 1930) and *Differential and Integral Calculus* (1934).[3] The former opens with two prefaces, one intended for the student and the other for the teacher. The preface for the student begins thus:

1. Please don't read the preface for the teacher.

2. I will ask of you only the ability to read English and to think logically—no high school mathematics, and certainly no higher mathematics.

3. Please forget everything you have learned in school; for you haven't learned it.

4. The multiplication table will not occur in this book, not even the theorem,

$$2 \cdot 2 = 4,$$

but I would recommend, as an exercise, that you define

$$2 = 1 + 1,$$
$$4 = (((1 + 1) + 1) + 1),$$

and then prove the theorem.

The preface for the teacher ends with these words:

My book is written, as befits such easy material, in merciless telegram style ("Axiom," "Definition," "Theorem," "Proof,'" occasionally "Preliminary Remark").... I hope that I have written this book in such a way that a normal student can read it in two days. And then (since he already knows the formal rules from school) he may forget its contents.

While it is not clear whom Landau may have considered a "normal student," it is hard to believe that an average student, or even a mathematics professor, could master in two days the 301 theorems of the book, written in almost hieroglyphic form in the book's 134 pages (fig. 91).

His textbook on the calculus is just 372 pages long—a far cry from today's thousand-page texts. Not a single illustration graces the book—after all, illustrations would imply that geometric concepts are being used, and geometry was *schmieröl*. Again the preface sets the tone for the entire work. First Landau refers to his *Grundlagen*, which he says received "tolerant and even some friendly reviews." He goes on to say that "a reader whose main interest lies in the applications of the calculus... should not make this book his choice." "My task," he says, is that of "bringing out into the open the definitions and theorems which are often implicitly assumed and which serve as the mortar when the whole structure is being built up with all the right floors in the right places."

True to its word, the book is structured in a terse definition-theorem-proof style, with an occasional example following a theorem. Definition 25 introduces the derivative of a function, followed immediately by two theorems: that differentiability implies continuity, and that there exist everywhere-continuous, nowhere-differentiable functions. This last theorem is due to the German mathematician Karl Theodor Wilhelm Weierstrass (1815–1897), whose lifelong goal was to rid analysis of any vestige of intuitiveness, and whose rigorous approach served as a model to Landau. Landau gives as an example the function

$$f(x) = \lim_{m \to \infty} \sum_{i=0}^{m} \{4^i x\}/4^i,$$

where $\{x\}$ is the distance of x to its nearest integer, and then proves that $f(x)$ is everywhere continuous but nowhere differentiable; the proof takes up nearly five pages.[4]

Of special interest to us here is the chapter on the trigonometric functions. It begins thus:

Theorem 248:

$$\sum_{m=0}^{\infty} \frac{(-1)^m}{(2m+1)!} x^{2m+1}$$

converges everywhere.

Theorem 280: *If* $f(1)$ *and* $f(1+1)$ *are defined, then*

$$\sum_{n=1}^{1+1} f(n) = f(1) \dotplus f(1+1).$$

Proof: By Theorems 278 and 277, we have

$$\sum_{n=1}^{1+1} f(n) = \sum_{n=1}^{1} f(n) \dotplus f(1+1) = f(1) \dotplus f(1+1).$$

Theorem 281: *If* $f(n)$ *is defined for* $n \leqq x + y$, *then*

$$\sum_{n=1}^{x+y} f(n) = \sum_{n=1}^{x} f(n) \dotplus \sum_{n=1}^{y} f(x+n).$$

Proof: Fix x, and let $\mathfrak{M}$ be the set of all y for which this holds.

I) If $f(n)$ is defined for $n \leqq x + 1$, then we have by Theorems 278 and 277 that

$$\sum_{n=1}^{x+1} f(n) = \sum_{n=1}^{x} f(n) \dotplus f(x+1) = \sum_{n=1}^{x} f(n) \dotplus \sum_{n=1}^{1} f(x+n).$$

Hence 1 belongs to $\mathfrak{M}$.

II) Let y belong to $\mathfrak{M}$. If $f(n)$ is defined for $n \leqq x + (y+1)$, then we have by Theorem 278 (applied to $x + y$ instead of x) that

$$\sum_{n=1}^{x+(y+1)} f(n) = \sum_{n=1}^{(x+y)+1} f(n) = \sum_{n=1}^{x+y} f(n) \dotplus f((x+y)+1)$$

$$= \left(\sum_{n=1}^{x} f(n) \dotplus \sum_{n=1}^{y} f(x+n) \right) \dotplus f(x+(y+1))$$

$$= \sum_{n=1}^{x} f(n) \dotplus \left(\sum_{n=1}^{y} f(x+n) \dotplus f(x+(y+1)) \right),$$

which by Theorem 278 (applied to y instead of x, and to $f(x + n)$ instead of $f(n)$) is

$$= \sum_{n=1}^{x} f(n) \dotplus \sum_{n=1}^{y+1} f(x+n).$$

Hence $y + 1$ belongs to $\mathfrak{M}$, and Theorem 281 is proved.

Theorem 282: *If* $f(n)$ *and* $g(n)$ *are defined for* $n \leqq x$, *then*

$$\sum_{n=1}^{x} (f(n) \dotplus g(n)) = \sum_{n=1}^{x} f(n) \dotplus \sum_{n=1}^{x} g(n).$$

Proof: Let $\mathfrak{M}$ be the set of all x for which this holds.

I) If $f(1)$ and $g(1)$ are defined, then

$$\sum_{n=1}^{1} (f(n) \dotplus g(n)) = f(1) \dotplus g(1) = \sum_{n=1}^{1} f(n) \dotplus \sum_{n=1}^{1} g(n).$$

Hence 1 belongs to $\mathfrak{M}$.

FIG. 91. A page from Edmund Landau's *Foundations of Analysis* (1930).

(This, of course, is the power series $x - x^3/3! + x^5/5! - + \cdots$).
This is followed by

Definition 59:

$$\sin x = \sum_{m=0}^{\infty} \frac{(-1)^m}{(2m+1)!} x^{2m+1}.$$

sin is to be read "sine."

After defining $\cos x$ in a similar manner there come several theorems establishing the familiar properties of these functions. Then,

Theorem 258:

$$\sin^2 x + \cos^2 x = 1.$$

Proof:

$$1 = \cos 0 = \cos(x - x) = \cos x \cos(-x) - \sin x \sin(-x)$$

$$= \cos^2 x + \sin^2 x.$$

Thus, out of the blue and without any mention by name, the most famous theorem of mathematics is introduced: the Pythagorean Theorem.[5]

Today, when textbooks fiercely compete with one another and must sell well in order to justify their publication, it is doubtful that Landau's texts would find a wide audience. In prewar European universities, however, higher education was the privilege of a very few. Moreover, a professor had total authority to teach his class the way he chose, including the choice of a textbook. Most professors followed no text at all but lectured from their own notes, and it was up to the student to supplement these notes with other material. In this atmosphere Landau's texts were held in high esteem as offering a true intellectual challenge to the serious student.

Notes and Sources

1. The theorem says that the average density of the prime numbers—the number of primes below a given integer x, divided by x—approaches $1/\ln x$ as $x \to \infty$. The theorem was first conjectured by Gauss in 1792, when he was fifteen years old. It was first proved in 1896 by Jacques Salomon Hadamard (1865–1963) of France and de la Vallée-Poussin (1866–1962) of Belgium, working independently.

2. Constance Reid, *Courant in Göttingen and New York: The Story of an Improbable Mathematician* (New York: Springer-Verlag, 1976),

pp. 25–26 and 126–127. See also "In Memory of Edmund Landau: Glimpses from the Panorama of Number Theory and Analysis," in *Edmund Landau: Collected Works*, ed. L. Mirsky et al. (Essen: Thales Verlag, 1985), pp. 25–50.

3. Both works appeared in English translation by Chelsea Publishing Company, New York: *Foundations of Analysis: The Arithmetic of Whole, Rational, Irrational and Complex Numbers*, trans. F. Steinhardt (1951), and *Differential and Integral Calculus*, trans. Melvin Hausner and Martin Davis (1950). The excerpts given above are from the English translations.

4. In his preface Landau defends this approach thus: "Some mathematicians may think it unorthodox to give as the second theorem after the definition of the derivative, Weierstrass' theorem... To them I would say that while there are very good mathematicians who have never learned any proof of that theorem, it can do the beginner no harm to learn the simplest proof to date right from his textbook."

5. The number π is introduced a few lines farther down as the smallest positive solution of the equation $\cos x = 0$. This "universal constant" is then denoted by π; there is no mention whatsoever of the numerical value of this constant, nor its relation to the circle.

15

Fourier's Theorem

Fourier, not being noble, could not enter the artillery,
although he was a second Newton.
—François Jean Dominique Arago

Trigonometry has come a long way since its inception more than two thousand years ago. But three developments, more than all others, stand out as having fundamentally changed the subject: Ptolemy's table of chords, which transformed trigonometry into a practical, computational science; De Moivre's theorem and Euler's formula $e^{ix} = \cos x + i \sin x$, which merged trigonometry with algebra and analysis; and Fourier's theorem, to which we devote this last chapter.

Jean Baptiste Joseph Fourier was born in Auxerre in north-central France on March 21, 1768. By the age of nine both his father and mother had died. Through the influence of some friends of the family, Fourier was admitted to a military school run by the Benedictine order, where he showed an early talent in mathematics. France has had a long tradition of producing great scientists who also served their country in the military: René Descartes (1596–1650), the soldier-turned-philosopher who invented analytic geometry; Gaspard Monge (1746–1818), who developed descriptive geometry and who in 1792 became minister of the marines; Jean Victor Poncelet (1788–1867), who wrote his great work on projective geometry while a prisoner of war in the aftermath of Napoleon's retreat from Moscow in 1812; and the two Carnots, the geometrician Lazar Nicolas Marguerite Carnot (1753–1823), who became one of France's great military leaders, and his son, the physicist Nicolas Léonard Sadi Carnot (1796–1832), who began his career as a military engineer and went on to lay the foundations of thermodynamics. Young Fourier wished to follow the tradition and become an artillery officer; but being of the wrong social class (his father was a tailor), he was only able to get a mathematics lectureship in the military school. This, however, did not deter him from getting involved in public life: he actively supported the French Revolution in 1789 and later was arrested for defending the victims of the terror, barely escaping the guillotine. In the end Fourier was rewarded for his activities and in 1795 was offered a professorship at the prestigious École Polytechnique in Paris, where Lagrange and Monge were also teaching.

In 1798 Emperor Napoleon Bonaparte launched his great military campaign in Egypt. A man of broad interests in the arts and sciences, Napoleon asked several prominent scholars to join him, among them Monge and Fourier. Fourier was appointed governor of southern Egypt and in that capacity organized the workshops of the French occupation forces. Following the French defeat at the hands of the British in 1801, he returned home and became prefect of the district of Grenoble. Among his administrative duties was the supervision of road construction and drainage projects, all of which he executed with great ability. And if that was not enough to keep him busy, he was also appointed secretary of the Institut d'Égypts, and in 1809 completed a major work on ancient Egypt, *Préface historique.*

One often marvels at the enormous range of activities of many eighteenth- and nineteenth-century scholars. At the very time Fourier was exercising his administrative duties, he was deeply engaged in his mathematical research. He worked in two seemingly unrelated fields: the theory of equations, and mathematical physics. When only sixteen, he found a new proof of Descartes' rule of signs about the number of positive and negative roots of a polynomial. His became the standard proof found in modern algebra texts. He began working on a book entitled *Analyse des équations déterminées*, in which he anticipated linear programming. However, Fourier died before completing this work (it was edited for publication in 1831 by his friend Louis Marie Henri Navier). He also pioneered dimensional analysis—the study of relations among physical quantities based on their dimensions.

But it is in mathematical physics that Fourier left his greatest mark. He was particularly interested in the manner in which heat flows from a region of high temperature to one of lower temperature. Newton had already studied this question and found that the rate of cooling (drop in temperature) of an object is proportional to the difference between its temperature and that of its surroundings. Newton's law of cooling, however, governs only the *temporal* rate of change of temperature, not its *spatial* rate of change, or gradient. This latter quantity depends on many factors: the heat conductivity of the object, its geometric shape, and the initial temperature distribution on its boundary. To deal with this problem one must use the analytic tools of the continuum, in particular partial differential equations (see p. 53). Fourier showed that to solve such an equation one must express the initial temperature distribution as a sum of infinitely many sine and cosine terms—a *trigonometric* or *Fourier series*. Fourier began work on this subject as early as 1807 and later expanded it in his major work, *Theorié analytique de la chaleur* (Analytic theory of heat, 1822), which became a model for some of the great nineteenth-century treatises on mathematical physics.

Fourier died tragically in Paris on May 16, 1830, after falling from a flight of stairs. Few portraits of him survive. A bust created in 1831 was destroyed in World War II. A second bust, erected in his hometown in 1849, was melted down by the German occupiers, who used the metal for armament; but the mayor of Auxerre got word of the impending disaster and managed to rescue two bas-reliefs of the bust, and fortunately these survived. In 1844 the archeologist Jacques Joseph Champollion-Figeac (brother of the Egyptologist Champollion mentioned in the Prologue) wrote Fourier's biography, entitled *Fourier, Napoleon et les cent jours*.[1]

In his work Fourier was guided as much by his sound grasp of physical principles as by purely mathematical considerations. His motto was: "Profound study of nature is the most fertile source of mathematical discoveries." This brought him biting criticism from such purists as Lagrange, Poisson, and Biot, who attacked his "lack of rigor"; one suspects, however, that political motivations and personal rivalry played a role as well. Ironically, Fourier's work in mathematical physics would later lead to one of the purest of all mathematical creations—Cantor's set theory.

✧   ✧   ✧

The basic idea behind Fourier's theorem is simple. We know that the functions $\cos x$ and $\sin x$ each have period 2π, the functions $\cos 2x$ and $\sin 2x$ have period $2\pi/2 = \pi$, and in general the functions $\cos nx$ and $\sin nx$ have period $2\pi/n$. But if we form any *linear combination* of these functions—that is, multiply each by a constant and add the results—the resulting function still has period 2π (fig. 92). This leads us to the following observation:

Let $f(x)$ be any "reasonably behaved" periodic function with period 2π; that is, $f(x + 2\pi) = f(x)$ for all x in its domain.[2] We

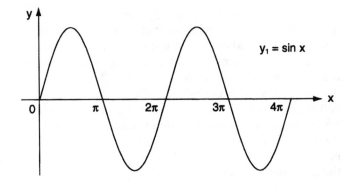

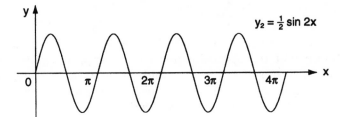

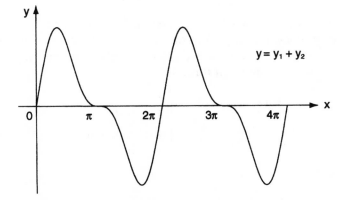

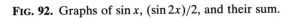

FIG. 92. Graphs of $\sin x$, $(\sin 2x)/2$, and their sum.

form the *finite* sum

$$
S_n(x) = \frac{a_0}{2} + a_1 \cos x + a_2 \cos 2x + a_3 \cos 3x + \cdots + a_n \cos nx
$$
$$
+ b_1 \sin x + b_2 \sin 2x + b_3 \sin 3x + \cdots + b_n \sin nx
$$
$$
= \frac{a_0}{2} + \sum_{m=1}^{n} (a_m \cos mx + b_m \sin mx), \tag{1}
$$

where the coefficients a_m and b_m are real numbers (the reason for dividing a_0 by 2 will become clear later); the subscript n under the $S(x)$ indicates that the sum depends on the number of sine and cosine terms present. Since $S_n(x)$ is the sum of terms of the form $\cos mx$ and $\sin mx$ for $m = 1, 2, 3, \ldots$, it is a periodic function of x with period 2π; the nature of this function, of course, depends on the coefficients a_m and b_m (as well as on n). We now ask: is it possible to determine these coefficients so that the sum (1), for large n, will approximate the *given* function $f(x)$ in the interval $-\pi < x < \pi$? In other words, can we determine the a_m's and b_m's so that

$$f(x) \approx \frac{a_0}{2} + \sum_{m=1}^{n}(a_m \cos mx + b_m \sin mx) \tag{2}$$

for every point in the interval $-\pi < x < \pi$? Of course, we require that the approximation should improve as n increases, and that for $n \to \infty$ it should become an equality; that is, $\lim_{n\to\infty} S_n(x) = f(x)$. If this indeed is possible, we say that the series (2) *converges* to $f(x)$ and write

$$f(x) = \frac{a_0}{2} + \sum_{m=1}^{\infty}(a_m \cos mx + b_m \sin mx). \tag{3}$$

In what follows we shall assume that the series (2) does indeed converge to $f(x)$ in the interval $-\pi < x < \pi$,[3] and we will show how to determine the coefficients.[4] Our starting point is the three integration formulas

$$\int_{-\pi}^{\pi} \sin mx \sin nx \, dx = \begin{cases} 0, & \text{if } m \neq n \\ \pi, & \text{if } m = n \neq 0 \end{cases}$$

$$\int_{-\pi}^{\pi} \cos mx \cos nx \, dx = \begin{cases} 0, & \text{if } m \neq n \\ \pi, & \text{if } m = n \neq 0 \end{cases}$$

and

$$\int_{-\pi}^{\pi} \sin mx \cos nx \, dx = 0 \quad \text{for all } m \text{ and } n,$$

known as the *orthogonality relations* for the sine and cosine (these formulas can be proved by using the product-to-sum formulas for each integrand and then integrating each term separately; note that when both m and n are zero, the integrand in the second formula is 1, so that we get $\int_{-\pi}^{\pi} dx = 2\pi$).

To find the coefficients a_m for $m = 1, 2, 3, \ldots$, we multiply equation (3) by $\cos mx$ and integrate it term-by-term over the interval $-\pi < x < \pi$.[5] In view of the orthogonality relations, all terms on the right side of the equation will be zero except the term $(a_m \cos mx) \cdot \cos mx = a_m \cos^2 mx = a_m(1 + \cos 2mx)/2$, whose integral from $-\pi$ to π is πa_m. We thus get

$$a_m = \frac{1}{\pi} \int_{-\pi}^{\pi} f(x) \cos mx \, dx, \quad m = 1, 2, 3, \cdots. \tag{4}$$

To find a_0 we repeat the process; but since we now have $m = 0$, multiplying equation (3) by $\cos 0x = 1$ leaves it unchanged, so we simply integrate it from $-\pi$ to π; again all terms will be zero except the term $(a_0/2) \int_{-\pi}^{\pi} dx = (a_0/2)(2\pi) = \pi a_0$. We thus get

$$a_0 = \frac{1}{\pi} \int_{-\pi}^{\pi} f(x) \, dx. \tag{5}$$

Note that equation (5) is actually a special case of equation (4) for $m = 0$; this is why we chose the constant term in equation (3) as $a_0/2$. Had we chosen it to be a_0, the right side of equation (5) would have to be divided by 2.

Finally, to get b_m we multiply equation (3) by $\sin mx$ and again integrate from $-\pi$ to π; the result is

$$b_m = \frac{1}{\pi} \int_{-\pi}^{\pi} f(x) \sin mx \, dx, \quad m = 0, 1, 2, \cdots. \tag{6}$$

Equations (4) through (6) are known as Euler's formulas (yes, two more formulas named after Euler!), and they allow us to find each coefficient of the Fourier series. Of course, depending on the nature of $f(x)$, the actual integration may or may not be executable in terms of the elementary functions; in the latter case we must resort to numerical integration.

Let us now apply this procedure to some simple functions. Consider the function $f(x) = x$, regarded as a periodic function over the interval $-\pi < x < \pi$; its graph has the saw-tooth shape shown in figure 93. Because this is an *odd* function (that is, $f(-x) = -f(x)$), the integrand in the first of Euler's equations is odd; and since the limits of integration are symmetric

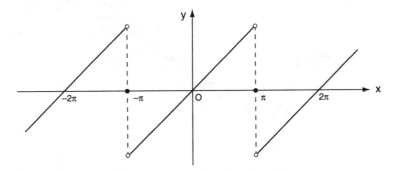

FIG. 93. Graph of the periodic function $f(x) = x$, $-\pi < x < \pi$.

with respect to the origin, the resulting integral will be zero for all $m = 0, 1, 2, \ldots$. Thus all the a_m's are zero, and our series will consist of sine terms only. For the b_m's we have

$$b_m = \frac{1}{\pi} \int_{-\pi}^{\pi} x \sin mx \, dx = \frac{2}{\pi} \int_{0}^{\pi} x \sin mx \, dx.$$

Integration by parts leads to

$$b_m = \frac{2(-1)^{m+1}}{m}.$$

We thus have

$$f(x) = 2 \left(\frac{\sin x}{1} - \frac{\sin 2x}{2} + \frac{\sin 3x}{3} - + \cdots \right). \tag{7}$$

Figure 94 shows the first four partial sums of this series; we clearly see how the sine waves pile up near $x = \pm\pi$, but it is not so obvious that the series actually converges to the saw-tooth graph of figure 93 for *every* point of the interval, including the points of discontinuity at $x = \pm n\pi$. Indeed, in Fourier's time the fact that an infinite sum of smooth sine waves may converge to a function whose graph is anything but smooth was met with a great deal of disbelief.[6] But so were Zeno's paradoxes two thousand years earlier! When it comes to infinite processes, we can always expect some surprises around the corner.

Since equation (7) holds for any value of x, let us put in it some specific values. For $x = \pi/2$ we get

$$\frac{\pi}{2} = 2 \left(1 - \frac{1}{3} + \frac{1}{5} - \frac{1}{7} + - \cdots \right).$$

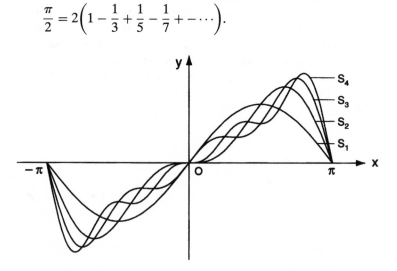

FIG. 94. First four partial sums of the Fourier expansion of $f(x) = x$, $-\pi < x < \pi$.

Dividing by 2 gives us the Gregory-Leibniz series (p. 159). For $x = \pi/4$ we get, after some labor,

$$\frac{\pi\sqrt{2}}{4} = 1 + \frac{1}{3} - \frac{1}{5} - \frac{1}{7} + + - - \cdots,$$

a little-known formula that connects the reciprocals of the odd integers with π and $\sqrt{2}$ (note that the right side of this series has the same terms as the Gregory-Leibniz series, but their signs alternate every two terms).

For the *even* function $f(x) = x^2$ (again regarded as a periodic function over the interval $-\pi < x < \pi$) we obtain, after twice integrating by parts, a Fourier series of cosine terms only:

$$f(x) = \frac{\pi^2}{3} - 4\left(\frac{\cos x}{1^2} - \frac{\cos 2x}{2^2} + \frac{\cos 3x}{3^2} - + \cdots\right). \qquad (8)$$

Substituting $x = \pi$ and simplifying results in

$$\frac{\pi^2}{6} = \frac{1}{1^2} + \frac{1}{2^2} + \frac{1}{3^2} + \cdots.$$

This is the famous formula that Euler discovered in 1734 in an entirely different and nonrigorous manner (see chapter 12). Many other series can be obtained in a similar way, as shown in figure 95.

✧   ✧   ✧

We have formulated Fourier's theorem for functions whose period is 2π, but it can easily be adjusted to functions with an arbitrary period P by the substitution $x' = (2\pi/P)x$. It then becomes more convenient to formulate the theorem in terms of the *angular frequency* ω (omega), defined as $\omega = 2\pi/P$. Fourier's theorem then says that any periodic function can be written as the sum of infinitely many sine and cosine terms whose angular frequencies are $\omega, 2\omega, 3\omega$, and so on. The lowest of these frequencies (i.e., ω) is the *fundamental frequency*, and its higher multiples are known as *harmonics*.

The word "harmonic," of course, comes to us from music, so let us digress for a moment into the world of sound. A *musical sound*—a tone—is produced by the regular, periodic vibrations of a material body such as a violin string or the air column of a flute. These regular vibrations produce in the ear a sense of pitch that can be written as a note on the musical staff. By contrast, nonmusical sounds—noises—are the result of irregular, random vibrations, and they generally lack a sense of pitch. Music, then, is the realm of periodic vibrations.[7]

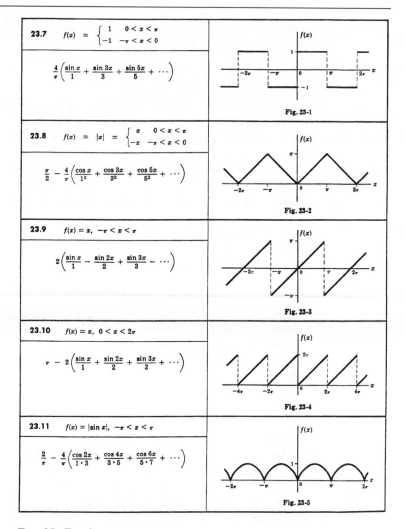

FIG. 95. Fourier expansion of some elementary functions.

The pitch of a musical sound is determined by the frequency of its vibrations: the higher the frequency, the higher the pitch. For example, the note C ("middle C" on the staff) corresponds to a frequency of 264 hertz, or cycles per second; the note A above C, to 440 hertz, and the note C′ one octave above C, to 528 hertz.[8] Musical *intervals* correspond to frequency *ratios*: an octave corresponds to the ratio 2 : 1, a fifth to 3 : 2, a fourth to 4 : 3, and so on (the names "octave," "fifth," and "fourth" derive from the positions of these intervals in the musical scale).

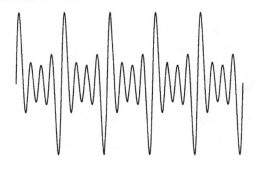

Fig. 96. Sound wave of a musical tone.

The simplest musical tone is a *pure tone*; it is produced by a sine wave, or—to use a term from physics—by *simple harmonic motion*.[9] A pure tone can be generated by an electronic synthesizer, but all natural musical instruments produce tones whose wave profiles, while periodic, are rather complicated (fig. 96). Nevertheless, these tones can always be broken down into their simple sine components—their *partial tones*—according to Fourier's theorem. Musical tones, then, are *compound tones*, whose constituent sine waves are the harmonics of the fundamental (lowest) frequency.[10]

The harmonics of a musical sound are not a mere mathematical abstraction: a trained ear can actually hear them. In fact, it is these harmonics that give a tone its characteristic "color"— its musical texture. The brilliant sound of a trumpet is due to its rich harmonic content; the sound of a flute is poor in harmonics, hence its mellow color (fig. 97). Each instrument has its characteristic *acoustic spectrum*—its signature of harmonic components. Amazingly, the human ear can split a compound tone into its component pure tones and hear each one of them separately, like a prism that splits white light into its rainbow colors

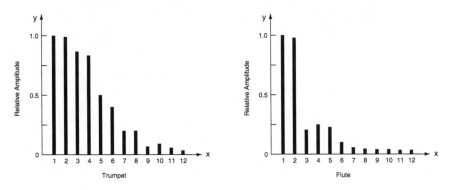

Fig. 97. Acoustic spectrum of a trumpet (left) and a flute (right).

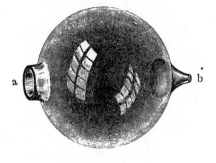

FIG. 98. Resonator.

(whence the name "spectrum"). The ear, in effect, is a Fourier analyzer.[11]

In the nineteenth century these ideas were new: scientists as well as musicians found it difficult to believe that a musical tone is actually the algebraic sum of all its harmonic components. The great German physicist and physiologist Herman Ludwig Ferdinand von Helmholtz (1821-1894) demonstrated the existence of partial tones by using *resonators*—small glass spheres of various sizes, each capable of enhancing one particular frequency in a compound tone (fig. 98). A series of these resonators formed a primitive Fourier analyzer analogous to the human ear. Helmholtz also did the reverse: by combining different simple tones of various frequencies and amplitudes, he was able to imitate the sound of actual musical instruments, anticipating the modern electronic synthesizer.

When we write the set of harmonics $1, 2, 3, \ldots$ in musical notation, we get the sequence of notes shown in figure 99. This sequence is known as the *harmonic series*, and it plays a crucial

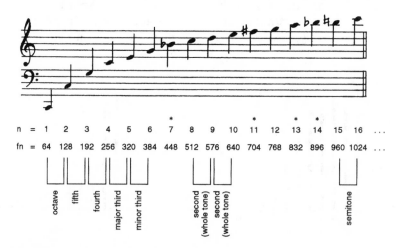

FIG. 99. The harmonic series.

role in musical theory: it is from this series that the fundamental musical intervals are derived.[12] That this series should have the same name as the mathematical series $1 + 1/2 + 1/3 + \cdots$ is no coincidence: the terms of the latter are precisely the periods of the harmonics in the former. Moreover, each term of the series $1 + 1/2 + 1/3 + \cdots$ is the *harmonic mean*[13] of the two terms immediately preceding and following it. These are just two examples of the numerous occurrences of the word "harmonic" in mathematics, reflecting the intimate ties that connect these two great creations of the human mind.

The importance of Fourier's theorem, of course, is not limited to music: it is at the heart of all periodic phenomena. Fourier himself extended the theorem to *nonperiodic* functions, regarding them as limiting cases of periodic functions whose period approaches infinity. The Fourier series is then replaced by an *integral* that represents a continuous distribution of sine waves over all frequencies. This idea proved of enormous importance to the development of quantum mechanics early in our century. The mathematics of Fourier's integral is more complicated than that of the series, but at its core are the same two functions that form the backbone of all trigonometry: the sine and cosine.[14]

NOTES AND SOURCES

1. There is no biography of Fourier in English. A brief sketch of his life can be found in Eric Temple Bell, *Men of Mathematics* (Harmondsworth, U.K.: Penguin Books, 1965), vol. 1, chap. 12. The biographical sketch of Fourier in this chapter is based in part on the article on Fourier by Jerome R. Ravetz and I. Grattan-Guiness in the *DSB*.

2. By "reasonably behaved" we mean that $f(x)$ is *sectionally smooth* on $-\pi < x < \pi$, i.e., that it is continuous and differentiable there except possibly at a finite number of finite jump discontinuities. At a jump discontinuity, we define $f(x)$ as $[f(x^-) + f(x^+)]/2$, that is, the mean between the values of $f(x)$ just to the left and right of the point in question. For a complete discussion, see Richard Courant, *Differential and Integral Calculus* (London: Blackie & Son, 1956), vol. 1, chap. 9.

3. Convergence is assured under the conditions stipulated in note 2.

4. The situation is somewhat analogous to the expansion of a function $f(x)$ in a power series $\sum_{i=0}^{n} a_i x^i$: we must determine the coefficients so that the sum will approximate the function at each point in the interval of convergence.

5. Term-by-term integration is permissible under the conditions mentioned in note 2.

6. For an interesting historical episode relating to this issue, see Paul J. Nahin, *The Science of Radio* (Woodbury, N.Y.: American Institute of Physics, 1995), pp. 85–86.

7. However, in our own era this traditional distinction has all but disappeared: witness the never-ending debate between classical music connoisseurs and rock fans as to what constitutes "real" music.

8. These frequencies are in accordance with the international standard known as *concert pitch*, in which A = 440 hertz. *Scientific pitch* is based on C = 256 hertz and has the advantage that all octaves of C correspond to powers of two; in this pitch A = 426.7 hertz.

9. The term "pure tone" refers both to a sine and a cosine vibration. This is because the human ear is not sensitive to the relative phase of a tone; that is, $\sin \omega t$ and $\sin (\omega t + \varepsilon)$ sound the same to the ear.

10. Strictly speaking there is a distinction between *overtones* in general—any set of higher frequencies present in a tone—and *harmonics*, those overtones whose frequencies are integral multiples of the fundamental frequency. Most musical instruments produce harmonic overtones, but some—notably drums and percussion—have nonharmonic components that cause their pitch to be less well defined.

11. By contrast, the eye does not have this capability: when blue and yellow light are superimposed, the result appears as green.

12. However, the presence of two different ratios, 9 : 8 and 10 : 9, for a whole tone causes difficulties when a melody is transposed (translated) from one scale to another. For this reason all modern instruments are tuned according to the *equal-tempered scale*, in which the octave consists of twelve equal semitones, each with the frequency ratio $(^{12}\sqrt{2}) : 1$. The numerical value of this ratio is 1.059, slightly less than the *just intonation* semitone 16 : 15 = 1.066. See my article, "What is there so Mathematical about Music?" *Mathematics Teacher*, September 1979, pp. 415–422.

13. The harmonic mean H of two positive numbers a and b is defind as $H = 2ab/(a + b)$. From this it follows that $1/H = (1/a + 1/b)/2$, i.e., the reciprocal of the harmonic mean is the arithmetic mean of the reciprocals of a and b. As an example, the harmonic mean of 1/2 and 1/4 is 1/3.

14. Fourier series have also been generalized to nontrigonometric functions, with appropriate orthogonality relations analogous to those for the sine and cosine. For details, see any text on advanced applied mathematics.

Appendixes

Appendix 1

Let's Revive an Old Idea

There are several ways to introduce the trigonometric functions: we can define them as ratios of sides in a right triangle, or in terms of the x- and y-coordinates of a point P on the unit circle, or as "wrapping functions" from the reals to some subset of the reals, or again as certain power series of the independent variable. Each approach has its merits, but clearly not all are equally suitable in the classroom. As I have mentioned in the Preface, the so-called "New Math" has imposed on trigonometry the language and formalities of abstract set theory—certainly not the best way to motivate the beginning student. I would like to suggest that we go back to an old idea: interpret the trigonometric functions as *projections*. To preempt criticism, let me say from the outset that nothing in this approach is new, but it reflects a shift in emphasis from the abstract to the practical. Let's not forget that trigonometry is, first and foremost, a *practical* discipline, born out of and deeply rooted in applications.

In figure 100, let $P(x, y)$ be a point on the unit circle, and let the angle between the positive x-axis and OP be θ (measured in degrees or radians). We define $\cos \theta$ and $\sin \theta$ as the *projections of OP on the x- and y-axes*, respectively. Since $OP = 1$, these projections are simply the x- and y-coordinates of the point P:

$$\cos \theta = OR = x, \quad \sin \theta = PR = y.$$

The tangent function is defined as $\tan \theta = y/x$, but this too can be viewed as a projection: referring again to fig. 100, draw the vertical tangent line to the unit circle at the point $S(1, 0)$, and call this line the t-axis. Extend OP until it meets the t-axis at Q. We have $\tan \theta = y/x = PR/OR$. But triangles OPR and OQS are similar, and therefore $PR/OR = QS/OS$. Recalling that $OS = 1$ and denoting the line segment QS by t, we have

$$\tan \theta = QS = t.$$

Thus $\tan \theta$ *is the projection of OP on the t-axis*. The function $\cot \theta$ can be similarly defined as the projection of OP on the horizontal tangent line to the circle at $T(0, 1)$ (fig. 101); we have

$$\cot \theta = QT = c.$$

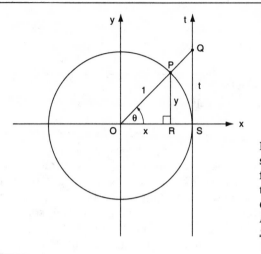

Fig. 100. The cosine, sine, and tangent functions as projections of the unit circle: $OR = x = \cos\theta$, $RP = y = \sin\theta$, $SQ = t = \tan\theta$.

So far the segments OR, PR, and QS were nondirected, but let us now think of them as directed line segments. This immediately leads us to conclude that $\tan\theta$ is positive for $0° < \theta < 90°$ (i.e., in quadrant I) and negative for $270° < \theta < 360°$ (quadrant IV). If θ is in quadrant II, we project OP *backward* until it meets the t-axis at Q (fig. 102); since triangles OPR and $OP'R'$ are congruent, we have $\tan\theta = PR/OR = P'R'/OR' = SQ/1 = t$, so that $\tan\theta$ is now the negative line segment SQ. If θ is in quadrant III, projecting OP backward gives us again a positive value for SQ; we have here a geometric demonstration of the identity $\tan(\theta + 180°) = \tan\theta$.

As for $\sec\theta$ and $\csc\theta$, they too can be interpreted (indeed defined) as projections: again let P be a point on the unit circle (fig. 103); draw the tangent line to the circle at P and extend it

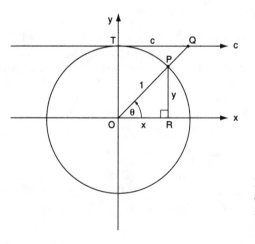

Fig. 101. The cotangent function as a projection: $TQ = c = \cot\theta$.

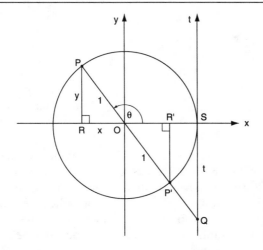

Fig. 102. The case when θ is in quadrant II.

until it meets the x- and y-axes at M and N, respectively. We have $\angle OPM = 90°$, so that triangles OPR and OMP are similar; hence $\sec \theta = 1/x = OP/OR = OM/OP = OM/1$, so that the line segment OM represents the value of $\sec \theta$ (again, it is a directed line segment, being negative when P is in quadrants II and III). Similarly, the line segment ON represents the value of $\csc \theta$. Moreover, since M and N always lie outside the circle, we see that the range of $\sec \theta$ and $\csc \theta$ is $(-\infty, -1] \cup [1, \infty)$.

Viewing all six trigonometric functions as projections allows us to see, quite literally, how these functions vary with θ: we only need to follow the various line segments as P moves around the circle. For example, to an observer watching from above, $\cos \theta$ would appear as the to-and-fro motion of the "shadow"

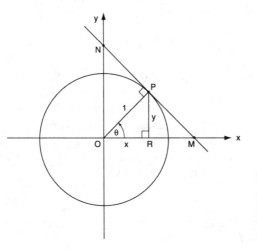

Fig. 103. The secant and cosecant functions as projections: $OM = 1/x = \sec \theta$, $ON = 1/y = \csc \theta$.

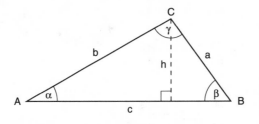

FIG. 104. The Law of Sines and the Law of Cosines as projections.

of P on the x-axis. This is even more dramatically illustrated in the case of $\tan\theta$: as θ increases, the shadow of P on the t-axis (point Q) rises slowly at first, than at an ever-increasing rate until it disappears at infinity as θ approaches 90°. Like the sweeping beam of a strobe light illuminating a dark wall, we have here a vivid graphic illustration of the peculiar behavior of $\tan\theta$ near its asymptotes.[1]

✧   ✧   ✧

The concept of projection can be put to good use not only in defining the trigonometric functions. Consider any triangle ABC (fig. 104). Drop the perpendicular h from C to AB. The projections of the sides a and b on h—let's call them the "vertical projections"—must of course be equal; hence

$$a\sin\beta = b\sin\alpha, \tag{1}$$

or $a/\sin\alpha = b/\sin\beta$. Repeating this argument for sides b and c, we get $b/\sin\beta = c/\sin\gamma$; thus $a/\sin\alpha = b/\sin\beta = c/\sin\gamma$, which is the Law of Sines.

On the other hand, the projections of a and b on AB (the "horizontal projections") must add up to the length of AB, that is, to c; we thus have

$$c = a\cos\beta + b\cos\alpha, \tag{2}$$

with similar equations for a and b (note that the formula is valid even if one angle, say α, is obtuse, in which case $\cos\alpha$ will be negative). It would be entirely appropriate to call equation (2) the Law of Cosines (a name usually given to the more familiar formula $c^2 = a^2 + b^2 - 2ab\cos\gamma$)—all the more so because it involves *two* cosines, thus justifying the plural "s." As an immediate consequence of equation (2) we have

$$c \leq a + b, \tag{3}$$

with equality if, and only if, $\alpha = \beta = 0°$. This is the famous *triangle inequality*.

Because it involves five variables—two angles and three sides—the usefulness of equation (2) for solving triangles is rather limited. We can, however, use equations (1) and (2) together to reduce the number of variables. Squaring equation (2), we have

$$c^2 = a^2 \cos^2 \beta + b^2 \cos^2 \alpha + 2ab \cos \alpha \cos \beta$$
$$= a^2(1 - \sin^2 \beta) + b^2(1 - \sin^2 \alpha) + 2ab \cos \alpha \cos \beta$$
$$= a^2 + b^2 - (a \sin \beta)(a \sin \beta) - (b \sin \alpha)(b \sin \alpha)$$
$$+ 2ab \cos \alpha \cos \beta.$$

Using equation (1), we can write this as

$$= a^2 + b^2 - (a \sin \beta)(b \sin \alpha)$$
$$- (a \sin \beta)(b \sin \alpha) + 2ab \cos \alpha \cos \beta$$
$$= a^2 + b^2 + 2ab(\cos \alpha \cos \beta - \sin \alpha \sin \beta)$$
$$= a^2 + b^2 + 2ab \cos (\alpha + \beta).$$

But $\cos (\alpha + \beta) = \cos (180° - \gamma) = - \cos \gamma$; we thus get

$$c^2 = a^2 + b^2 - 2ab \cos \gamma, \tag{4}$$

which is the familiar form of the Law of Cosines. Thus the sine and the cosine laws simply express the fact that in a triangle, the perpendicular from any vertex to the opposite base is the vertical projection of either of the adjacent sides, and the base is the sum of their horizontal projections.

NOTE

1. Regrettably, many textbooks plot the graphs of $\tan \theta$ and $\cot \theta$ based on just a few scattered values of θ (in one case, as few as three!), arbitrarily chosen between $-90°$ and $90°$. Surely no real understanding of the peculiar behavior of these functions can be gained this way.

Appendix 2

Barrow's Integration of $\sec\phi$

We give here in modern notation Isaac Barrow's proof (1670) that $\int_0^\phi \sec t\, dt = \ln\tan(45° + \phi/2)$ (see p. 178). This proof marks the first use of the technique of decomposition into partial fractions.[1]

We begin with

$$\sec\phi = \frac{1}{\cos\phi} = \frac{\cos\phi}{\cos^2\phi}$$

$$= \frac{\cos\phi}{1 - \sin^2\phi} = \frac{\cos\phi}{(1+\sin\phi)(1-\sin\phi)}$$

$$= \frac{1}{2}\left(\frac{\cos\phi}{1+\sin\phi} + \frac{\cos\phi}{1-\sin\phi}\right).$$

Therefore,

$$\int \sec\phi\, d\phi = \frac{1}{2}\int \left(\frac{\cos\phi}{1+\sin\phi} + \frac{\cos\phi}{1-\sin\phi}\right) d\phi.$$

The first term inside the integral is of the form u'/u, where u is a function of ϕ, and the second term is of the form $-u'/u$; using the formula $\int (u'/u)\, d\phi = \ln|u(\phi)| + C$, we have

$$\int \sec\phi\, d\phi = \frac{1}{2}[\ln|1 + \sin\phi| - \ln|1 - \sin\phi|] + C.$$

Using the familiar division property of logarithms, this becomes

$$= \frac{1}{2}\ln\left|\frac{1+\sin\phi}{1-\sin\phi}\right| + C.$$

We multiply and divide the expression inside the logarithm by $(1+\sin\phi)$; in the denominator we get $(1-\sin\phi)(1+\sin\phi) = 1 - \sin^2\phi = \cos^2\phi$, and so

$$\int \sec\phi\, d\phi = \frac{1}{2}\ln\frac{(1+\sin\phi)^2}{\cos^2\phi} + C.$$

Using the power property of logarithms, this becomes

$$= \ln\left|\frac{1+\sin\phi}{\cos\phi}\right| + C.$$

Writing $\phi = 2(\phi/2)$ and using the double-angle formulas for sine and cosine, we get

$$= \ln \left| \frac{1 + 2\sin \phi/2 \cos \phi/2}{\cos^2 \phi/2 - \sin^2 \phi/2} \right| + C$$

$$= \ln \left| \frac{(\cos \phi/2 + \sin \phi/2)^2}{(\cos \phi/2 + \sin \phi/2)(\cos \phi/2 - \sin \phi/2)} \right| + C$$

$$= \ln \left| \frac{\cos \phi/2 + \sin \phi/2}{\cos \phi/2 - \sin \phi/2} \right| + C.$$

Finally, dividing the numerator and denominator of the expression inside the logarithm by $\cos \phi/2$, we get

$$= \ln \left| \frac{1 + \tan \phi/2}{1 - \tan \phi/2} \right| + C$$

$$= \ln \left| \tan \left(\frac{\pi}{4} + \frac{\phi}{2} \right) \right| + C.$$

Turning now to the definite integral, we have

$$\int_0^\phi \sec t \, dt = \ln \left| \tan \left(\frac{\pi}{4} + \frac{\phi}{2} \right) \right| - \ln \tan \frac{\pi}{4}.$$

But $\ln \tan \pi/4 = \ln 1 = 0$, so we finally have

$$\int_0^\phi \sec t \, dt = \ln \tan \left(\frac{\pi}{4} + \frac{\phi}{2} \right)$$

(we have dropped the absolute value sign because in the relevant range of ϕ, namely $-\pi/2 < \phi < \pi/2$, $\tan (\pi/4 + \phi/2)$ is positive).

Today one solves this integral by the substitution $u = \tan t/2$, $du = [(1/2) \sec^2 t/2] \, dt$, but it is still a tough nut to crack for beginning calculus students.

SOURCE

1. This derivation is based on the article, "An Application of Geography to Mathematics: History of the Integral of the Secant" by V. Frederick Rickey and Philip M. Tuchinsky, in the *Mathematics Magazine*, vol. 53, no. 3 (May 1980).

Appendix 3

Some Trigonometric Gems

"Beauty is in the eye of the beholder," says an old proverb. I have collected here a sample of trigonometric formulas that will appeal to anyone's sense of beauty. Some of these formulas are easy to prove, others will require some effort on the reader's behalf. My selection is entirely subjective: trigonometry abounds in beautiful formulas, and no doubt the reader can find many others that are equally appealing.

1. FINITE FORMULAS

$$\sin^2 \alpha + \cos^2 \alpha = 1$$

$$\sin^4 \alpha - \cos^4 \alpha = \sin^2 \alpha - \cos^2 \alpha$$

$$\sec^2 \alpha + \csc^2 \alpha = \sec^2 \alpha \csc^2 \alpha$$

$$\sin(\alpha + \beta)\sin(\alpha - \beta) = \sin^2 \alpha - \sin^2 \beta$$

$$\tan(45° + \alpha)\tan(45° - \alpha) = \cot(45° + \alpha)\cot(45° - \alpha) = 1$$

$$\sin(\alpha + \beta + \gamma) + \sin \alpha \sin \beta \sin \gamma$$
$$= \sin \alpha \cos \beta \cos \gamma + \sin \beta \cos \gamma \cos \alpha + \sin \gamma \cos \alpha \cos \beta$$

Let $f(\alpha, \beta) = \cos^2 \alpha + \sin^2 \alpha \cos 2\beta$;

   then $f(\alpha, \beta) = f(\beta, \alpha)$.

Let $g(\alpha, \beta) = \sin^2 \alpha - \cos^2 \alpha \cos 2\beta$;

   then $g(\alpha, \beta) = g(\beta, \alpha)$.

In the following relations, let $\alpha + \beta + \gamma = 180°$:

$$\sin \alpha + \sin \beta + \sin \gamma = 4 \cos \alpha/2 \cos \beta/2 \cos \gamma/2$$

$$\sin 2\alpha + \sin 2\beta + \sin 2\gamma = 4 \sin \alpha \sin \beta \sin \gamma$$

$$\sin 3\alpha + \sin 3\beta + \sin 3\gamma = -4 \cos 3\alpha/2 \cos 3\beta/2 \cos 3\gamma/2$$

$$\cos \alpha + \cos \beta + \cos \gamma = 1 + 4 \sin \alpha/2 \sin \beta/2 \sin \gamma/2$$

$$\cos^2 2\alpha + \cos^2 2\beta + \cos^2 2\gamma - 2\cos 2\alpha \cos 2\beta \cos 2\gamma = 1$$

$$\tan \alpha + \tan \beta + \tan \gamma = \tan \alpha \tan \beta \tan \gamma\ ^1$$

$$0 < \sin \alpha + \sin \beta + \sin \gamma \le (3\sqrt{3})/2,$$

with equality if, and only if, $\alpha = \beta = \gamma = 60°$.

In any acute triangle,

$$\tan \alpha + \tan \beta + \tan \gamma \ge 3\sqrt{3},$$

with equality if, and only if, $\alpha = \beta = \gamma = 60°$.

In any obtuse triangle,

$$-\infty < \tan \alpha + \tan \beta + \tan \gamma < 0.$$

2. INFINITE FORMULAS[2]

$$\sin x = x - x^3/3! + x^5/5! - + \cdots$$

$$\cos x = 1 - x^2/2! + x^4/4! - + \cdots$$

$$\sin x = x(1 - x^2/\pi^2)(1 - x^2/4\pi^2)(1 - x^2/9\pi^2) \cdots$$

$$\cos x = (1 - 4x^2/\pi^2)(1 - 4x^2/9\pi^2)(1 - 4x^2/25\pi^2) \cdots$$

$$\tan x = 8x[1/(\pi^2 - 4x^2) + 1/(9\pi^2 - 4x^2)$$
$$+ 1/(25\pi^2 - 4x^2) + \cdots]$$

$$\sec x = 4\pi[1/(\pi^2 - 4x^2) - 3/(9\pi^2 - 4x^2)$$
$$+ 5/(25\pi^2 - 4x^2) - + \cdots]$$

$$(\sin x)/x = \cos x/2 \cos x/4 \cos x/8 \cdots$$

$$(1/4)\tan \pi/4 + (1/8)\tan \pi/8 + (1/16)\tan \pi/16 + \cdots = 1/\pi$$

$$\tan^{-1} x = x - x^3/3 + x^5/5 - + \cdots, \qquad -1 < x < 1.$$

Notes

1. The companion formula

$$\cot \alpha + \cot \beta + \cot \gamma = \cot \alpha \cot \beta \cot \gamma$$

holds true only for $\alpha + \beta + \gamma = 90°$.

2. For a sample of Fourier series, see figure 95, p. 206. Numerous other trigonometric series can be found in *Summation of Series*, collected by L. B. W. Jolley (1925; rpt. New York: Dover, 1961), chaps. 14 and 16.

Appendix 4

Some Special Values of $\sin \alpha$

$$\sin 0° = 0 = \frac{\sqrt{0}}{2}, \quad \sin 30° = \frac{1}{2} = \frac{\sqrt{1}}{2}, \quad \sin 45° = \frac{\sqrt{2}}{2},$$

$$\sin 60° = \frac{\sqrt{3}}{2}, \quad \sin 90° = 1 = \frac{\sqrt{4}}{2}.$$

$$\sin 15° = \frac{\sqrt{2 - \sqrt{3}}}{2} = \frac{\sqrt{6} - \sqrt{2}}{4},$$

$$\sin 75° = \frac{\sqrt{2 + \sqrt{3}}}{2} = \frac{\sqrt{6} + \sqrt{2}}{4}.$$

$$\sin 18° = \frac{-1 + \sqrt{5}}{4},$$

$$\sin 36° = \frac{\sqrt{10 - 2\sqrt{5}}}{4},$$

$$\sin 54° = \frac{1 + \sqrt{5}}{4},$$

$$\sin 72° = \frac{\sqrt{10 + 2\sqrt{5}}}{4}.$$

The last four values are related to a regular pentagon. For example, the side of a regular pentagon inscribed in a unit circle is $2 \sin 36°$, its diagonal is $2 \sin 72°$, and their ratio is $2 \sin 54°$. These values are also related to the "golden section": the ratio in which a line segment must be divided if the whole segment is to the longer part as the longer part is to the shorter. This ratio, denoted by ϕ, is equal to $(1 + \sqrt{5})/2 \approx 1.618$, that is, to $2 \sin 54°$.

Repeated use of the half-angle formula for the sine leads to the following expressions, where $n = 1, 2, 3, \ldots$:

$$\sin \frac{45°}{2^n} = \frac{\sqrt{2 - \sqrt{2 + \sqrt{2 + \cdots + \sqrt{2}}}}}{2}$$

$$(n + 1 \text{ nested square roots})$$

$$\sin \frac{15°}{2^n} = \frac{\sqrt{2 - \sqrt{2 + \sqrt{2 + \cdots + \sqrt{3}}}}}{2}$$

$(n + 2$ nested roots$)$

$$\sin \frac{18°}{2^n} = \frac{\sqrt{8 - 2\sqrt{8 + 2\sqrt{8 + \cdots + 2\sqrt{10 + 2\sqrt{5}}}}}}{4}$$

$(n + 2$ nested roots$)$

Bibliography

Aaboe, Asger. *Episodes from the Early History of Mathematics.* New York: Random House, 1964.

Ball, W. W. Rouse. *A Short Account of the History of Mathematics.* 1908. Rpt. New York: Dover, 1960.

Beckman, Petr. *A History of π.* Boulder, Colo.: Golem Press, 1977.

Bell, Eric Temple. *Men of Mathematics.* 2 vols. 1937. Rpt. Harmondsworth, U.K.: Penguin Books, 1965.

_____. *The Development of Mathematics.* 1945. 2d ed. Rpt. New York: Dover, 1992.

Berthon, Simon, and Andrew Robinson. *The Shape of the World: The Mapping and Discovery of the Earth.* Chicago: Rand McNally, 1991.

Bond, John David. "The Development of Trigonometric Methods down to the Close of the XVth Century," *Isis* 4 (October 1921), pp. 295–323.

Boyer, Carl B. *A History of Mathematics.* 1968. Rev. ed. New York: John Wiley, 1989.

Braunmühl, Anton von. *Vorlesungen über die Geschichte der Trigonometrie.* 2 vols. Leipzig: Teubner, 1900–1903.

Brown, Lloyd A. *The Story of Maps.* 1949. Rpt. New York: Dover, 1979.

Burton, David M. *The History of Mathematics: An Introduction.* Boston: Allyn and Bacon, 1985.

Cajori, Florian. *A History of Mathematics.* 1893. 2d ed. New York: Macmillan, 1919.

_____. *A History of Mathematical Notations.* Vol. 2: *Higher Mathematics.* 1929. Rpt. Chicago: Open Court, 1952.

_____. *William Oughtred: A Great Seventeenth-Century Teacher of Mathematics.* Chicago: Open Court, 1916.

Chase, Arnold Buffum. *The Rhind Mathematical Papyrus.* 1927–1929. Rpt. Reston, Virginia: National Council of Teachers of Mathematics, 1979.

Courant, Richard. *Differential and Integral Calculus.* 2 vols. 1934. Rpt. London: Blackie & Son, 1956.

Dantzig, Tobias. *The Bequest of the Greeks.* New York: Charles Scribner's Sons, 1955.

Dörrie, Heinrich. *100 Great Problems of Elementary Mathematics: Their History and Solution.* Trans. David Anin. 1958. Rpt. New York: Dover, 1965.

Dunham, William. *Journey through Genius: The Great Theorems of Mathematics.* New York: John Wiley, 1990.

Euclid. *The Thirteen Books of Euclid's Elements.* 3 vols. Trans. from the text of Heiberg with introduction and commentary by Sir Thomas Heath. New York, 1956.

Eves, Howard. *An Introduction to the History of Mathematics.* 1964. Rpt. Philadelphia: Saunders College Publishing, 1983.

Gheverghese, George Joseph. *The Crest of the Peacock: Non-European Roots of Mathematics.* Harmondsworth, U.K.: Penguin Books, 1991.

Gillings, Richard J. *Mathematics in the Time of the Pharaohs.* 1972. Rpt. New York: Dover, 1982.

Gillispie, Charles Coulston, ed. *Dictionary of Scientific Biography.* 16 vols. New York: Charles Scribner's Sons, 1970–1980.

Helden, Albert van. *Measuring the Universe: Cosmic Dimensions from Aristarchus to Halley.* Chicago: University of Chicago Press, 1985.

Helmholtz, Hermann Ludwig Ferdinand von. *Sensations of Tone.* 1885. Trans. Alexander J. Ellis. New York: Dover, 1954.

Hollingdale, Stuart. *Makers of Mathematics.* Harmondsworth, U.K.: Penguin Books, 1989.

Jolley, L.B.W. *Summation of Series.* 1925. Rpt. New York: Dover, 1961.

Karpinski, Louis C. "Bibliographical Check List of All Works on Trigonometry Published up to 1700 A.D.," *Scripta Mathematica* 12 (1946), pp. 267–283.

Katz, Victor J. *A History of Mathematics: An Introduction.* New York: HarperCollins, 1993.

Klein, Felix. *Elementary Mathematics from an Advanced Standpoint.* Vol. 1: *Arithmetic, Algebra, Analysis.* 1924. Trans. E. R. Hedrick and C. A. Noble. Rpt. New York: Dover (no date).

Kline, Morris. *Mathematical Thought from Ancient to Modern Times.* 3 vols. New York: Oxford University Press, 1990.

Knopp, Konrad. *Elements of the Theory of Functions.* Trans. Frederick Bagemihl. New York: Dover, 1952.

Kramer, Edna E. *The Nature and Growth of Modern Mathematics.* 1970. Rpt. Princeton, N.J.: Princeton University Press, 1981.

Loomis, Elisha Scott. *The Pythagorean Proposition.* 1940. Rpt. Washington, D.C.: National Council of Teachers of Mathematics, 1972.

Maor, Eli. *e: The Story of a Number.* Princeton, N.J.: Princeton University Press, 1994.

Müller, Johann (Regiomontanus). *De triangulis omnimondis.* Trans. Barnabas Hughes with an Introduction and Notes. Madison, Wis.: University of Wisconsin Press, 1967.

Pedoe, Dan. *Geometry and the Liberal Arts.* New York: St. Martin's, 1976.

Simmons, George F. *Calculus with Analytic Geometry.* New York: McGraw-Hill, 1985.

Smith, David Eugene. *History of Mathematics.* Vol. 1: *General Survey of the History of Elementary Mathematics.* Vol. 2: *Special Topics of Elementary Mathematics.* 1923–1925. Rpt. New York: Dover, 1958.

Snyder, John P. *Flattening the Earth: Two Thousand Years of Map Projections.* Chicago: University of Chicago Press, 1993.

Struik, D. J., ed. *A Source Book in Mathematics, 1200–1800.* Cambridge, Mass.: Harvard University Press, 1969.

Taylor, C. A. *The Physics of Musical Sounds.* London: English Universities Press, 1965.

van der Werden, Bartel L. *Science Awakening: Egyptian, Babylonian and Greek Mathematics.* 1954. Trans. Arnold Dresden. 1961. Rpt. New York: John Wiley, 1963.

Wilford, John Noble. *The Mapmakers.* New York: Alfred A. Knopf, 1981.

Yates, Robert C. *Curves and Their Properties.* 1952. Rpt. Reston, Va.: National Council of Teachers of Mathematics, 1974.

Zeller, Mary Claudia. *The Development of Trigonometry from Regiomontanus to Pitiscus.* Ann Arbor, Mich.: Edwards Bros., 1944.

Credits for Illustrations

Title page and figs. 1 and 3: Courtesy of the National Council of Teachers of Mathematics. Reprinted from Arnold Buffum Chase, *Rhind Mathematical Papyrus*, 1979, with permission.

Fig. 6: ©1994 Carol Wright Gifts. Used by permission.

Fig. 12: Courtesy of the Mathematical Association of America. Used with permission.

Fig. 39: SPIROGRAPH® is a trademark of Hasbro, Inc. ©1997 Hasbro, Inc. Used with permission.

Fig. 91: Courtesy of Chelsea Publishing Company. Used with permission.

Fig. 95: Courtesy of the McGraw-Hill Companies. Reprinted from Murray R. Spiegel, *Mathematical Handbook of Formulas and Tables*, Schaum's Outline Series, 1968. Used with permission.

Index